DATE DUE			

JAMES WATT AND THE STEAM REVOLUTION

JAMES WATT
and the steam revolution

A DOCUMENTARY HISTORY BY ERIC ROBINSON AND A. E. MUSSON

AUGUSTUS M. KELLEY · PUBLISHERS
NEW YORK 1969

Published in the United States by

Augustus M. Kelley · Publishers

New York, New York 10001

© 1969, Eric Robinson and A. E. Musson

SBN 678 07756 8
Library of Congress Catalog Card No. 71–96795
Printed in Great Britain by Billing & Sons Limited

Contents

List of Illustrations

vii

Foreword

THIS VOLUME *of documents about James Watt and the steam revolution has come into existence largely through the generosity of Major David Gibson-Watt, M.C., M.P., who has not only allowed the editors to publish a further selection of his family papers but has also been his usual hospitable self. We are also grateful to the City of Birmingham Library Committee for permission to publish material from the Boulton and Watt Collection in the Birmingham Public Library, and to Mr. Arthur Westwood, the Assay Master of the Birmingham Assay Office, for similar permission to publish documents from the Tew MSS. Mr. Westwood has also allowed us to reproduce a photograph of the Peter Rouw wax medallion of Boulton which is in the Assay Office collection. Some of our other illustrations come from photographs taken by Mr. Belsher of the Birmingham Museum and Art Gallery for the Lunar Society Bicentenary exhibition. Mr. A. Gunstone of the Museum was kind enough to draw our attention to the medal of Watt depicted in this volume. To Mr. John Lowe, formerly Director of the Museum, we are indebted for many kindnesses, as we are also to our friends of the Birmingham Reference Library who, with their customary efficiency, provided all the necessary facilities for research. Last but not least we would like to thank our typist, Miss Patricia Horne.*

This selection of documents, produced on the occasion of the bicentenary of Watt's first patent, comprises a good deal of hitherto unpublished material, and even where there has been earlier publication we have referred to the original documents. In the case of manuscripts, we have always printed the last-preferred wording by the writer, without indicating deletions and alterations. To have included these would have been to produce documents that would have been in some cases extremely difficult to follow, without providing additional information of much consequence except to the antiquarian. For the same reason we have expanded all contractions to make for easier reading and have corrected punctuation where necessary.

The colour plates in this volume were part of the illustrative material prepared for the Case of Boulton and Watt v. Bull *and correspond, apart from some very small differences of lettering, to the drawings submitted as part of the specifications for the various patents. Similar drawings do not exist, however, for the 1785 patent which was not relevant to this case, and we have therefore reproduced the plates from J. P. Muirhead's* Mechanical Inventions of James Watt.

In the body of our Introduction we have inserted in square brackets numbers which refer to the numbering of the documents in the List of Contents so that the reader may make cross-reference if he wishes to the documents under discussion. In addition, however, we have added prefatory remarks to most documents in order that the reader may see the reasons for our selection and may have some understanding of the context in which they may best be understood.

Despite the important pioneer work of J. P. Muirhead a vast number of letters and other papers relating to Britain's greatest engineer remains unpublished. A related collection of documents may be studied in Eric Robinson and Douglas McKie, Partners in Science: James Watt and Joseph Black, *which includes the correspondence of James Watt, Joseph Black and John Robison; but the urgent need still remains for a modern collected edition of Watt's papers similar to the project now in progress for the publication of the papers of Sir Isaac Newton.*

The present volume, it is hoped, may serve to suggest some of the riches yet untapped.

Introduction

The late H. W. Dickinson's account of James Watt is entitled *James Watt, Craftsman and Engineer* (Cambridge, 1936) and an explanation of the choice of title is offered early in the book where the author states:

> 'Even today, when the amount of knowledge to be acquired is so great and the years of training to be undergone are so many before a man is qualified to be an engineer, he must, in this country at any rate, have had practical experience in the field or in the workshop before he is considered fully trained. Thus the engineer is a craftsman and nearly everything that we see round us in the material world is rooted in craftsmanship.'[1]

Dr. Dickinson appears to have chosen to put 'craftsman' before 'engineer' in order to present not only the simple chronological sequence of James Watt's transition from instrument-maker to engineer, but also what seemed to him the inevitable logical progression, as from chrysalis to butterfly, through which a man had to pass before emerging in full splendour as an engineer proper. The logic of that progression at first sight seems inescapable in heavy engineering. Though it may be possible to imagine, in the twentieth century, some clean-handed white-collared intellectual passing directly from the university laboratory to the managerial offices of an engineering firm, in the eighteenth-century setting of coal-burning steam-engines producing a convulsive motion in beams made of immense tree-trunks, operating at the pit-head of some remote Cornish copper mine or alongside some stretch of desolate canal, surely what was required was a man who knew how to get grease on his hands. For eighteenth-century engineering the image of 'the man in the white suit' appears peculiarly inappropriate, and the practical man, the millwright, the horny-handed craftsman, seems destined by nature for the role of the engineer. Moreover, Samuel Smiles, in his *Lives of the Engineers*, that great classic in the history of British technology, seems dedicated to the same point of view:

> 'One of the most remarkable things about Engineering in England is, that its principal achievements have been accomplished, not by natural philosophers nor by mathematicians, but by men of humble station, for the most part self-educated. The educated classes of the last century [i.e. the eighteenth century] regarded with contempt mechanical men and mechanical subjects. . . . At a time when the Court, the Camp, and the Church formed the principal occupations of the higher classes, engineering was thought unscientific and ungenteel. . . . Nor did any of the great mechanics, who have since invented tools, engines, and machines, at all belong to the educated classes. They received no college education. Some of them could scarcely write their own names. These men gathered their practical knowledge in the workshop, or acquired it in manual labour. They rose to celebrity, mostly by their habits of observation, their powers of discrimination, their constant self-improvement, and their patient industry.'[2]

How much more dramatic when men rise to glory from such obscure origins, how much further must the Wheel of Fortune have turned, how much more abounding is the grace when it is conferred upon the chief of sinners! It is a natural human feeling to marvel at such changes: 'Ah, Bottom, how art thou translated!'

To what extent does the greatest engineer of the eighteenth century fit into this context?

[1] p. 14.
[2] S. Smiles, *Lives of the Engineers* (1874), *Early Engineering*, pp. xvi–xvii.

From his earliest days Watt was given to building model cranes and to working with his hands and it was for this reason he seems to have elected to become an instrument-maker. His father was described as a wright, and the son grew up among the bustling activity of dock-yards on the Clyde. Throughout his life Watt was the marvel of his friends for his ability to wrestle with the mysteries of handicraft, his garret workshop at Heathfield was packed with tools, and we know that as late as

'his eightieth year, he instructed a certain great bookseller always to have his boxes for books *planed on the inside:* [and] wrote to Sir Henry Raeburn (2nd March 1816) "I beg to advise your packer always *to rub his wood-screws on some candle-grease* before he screws them in: it will tend much to his own ease as well as to that of him who unscrews them".'[1]

This seems to be the typical craftsman of whom Smiles speaks. We know also that Watt constructed common fire engines at Kinneil for Dr. Roebuck and

'had a suit of pit cloaths made, in which [he] went down the shaft and like any of the other workmen changed the buckets of the pumps, and overhauled the engine, and continued to do so, till [he] was completely master of the practical workings of it'. [4; p. 41]

We know that in Cornwall he laboured alongside his engine-erectors in pouring rain to remedy the defects of his early engines, and that he reckoned never to ask one of his men to do a job that he was incapable of performing himself.

Yet this is not the whole of the story, and if one were to rest content with it as it stands a serious distortion would ensue. Professor Robison's 'Narrative' [2; p. 24] illustrates the point:

'I saw a Workman and expected no more – but was surprized to find a philosopher, as young as myself; and always ready to instruct me. I had the vanity to think myself a pretty good proficient in my favourite Study and was rather mortifyd at finding Mr. Watt so much my superior.'

And again:

'Every thing became to him the beginning of a new and serious study – Every thing became Science in his hands. . . .'

Robison's account of his scientific and mathematical discussions with Watt, of Watt's constant intellectual curiosity and scientific experimentation, the intimate friendships with men of science which Watt enjoyed from his earliest days at Glasgow, all testify to the fact that Watt was no mere craftsman, and he certainly did not consider himself in that light.

He was from the first deeply interested in mathematics, as we learn from Miss Jane Campbell's information [1], and in later life carefully superintended his own son's mathematical education, setting him problems from the same mathematics textbook that he himself used when a boy.

'I hope by this time you have got a master in mathematicks and that you apply yourself principally to such branches as you have not learnt before, or not thoroughly learnt. in plain or solid

[1] J. P. Muirhead, *The Origin and Progress of the Mechanical Inventions of James Watt* (1854), 3 vols., Introductory Memoir, Vol. I, pp. xix–xx.

2

geometry it will be sufficient that you ask his advice or explanation of any difficulty but in conic sections the doctrines of the sphere etc. you should go on regularly with him till you understand them. You should divide your lesson into 2 parts one to be dedicated to geometry and the other to the higher parts of Algebra such as cubic and Biquadratic Equations, the summation of series the squaring of curves and other applications of Algebra to geometry. . .'[1]

On 15 September 1781 Watt wrote to his relation, Mrs. Campbell, giving her advice about the education of her son.[2] He stressed the importance of the boy being well educated in mathematics to prepare him for a variety of professions, including that of engineer:

'Let him write well and be *thoroughly* taught arithmetic and algebra and too much pains cannot be bestowed on Geometry – Cultivate his taste for drawing in India ink but go no further least it degenerate into painting.'

In his 'Points necessary to be known by a Steam Engineer' [44] he lists first, 'The Laws of Mechanics as a Science' and only secondly 'Their usual practical Applications to the Construction of Machinery . . .' He requires the engineer to have an acquaintance with the laws of hydraulics and hydrostatics, with the doctrine of heat and cold and with the bulk of liquid under different pressures. He sees himself differentiated from the mass of working engineers by his constant attention to the principles of his subject and it is clear that his contemporaries thought of him in the same way. From the very first Boulton exerted an attraction upon Watt because he could offer to manufacture Watt's engines upon scientific principles different from the hit-and-miss methods of much eighteenth-century engineering [9].

Though this volume is restricted to documents about Watt and the steam-engine, we must guard against any interpretation of Watt's career which neglects the remarkable diversity of his scientific interests. The instrument-maker who built an organ *by the book* and in doing so not only mastered the theory of harmonics but also made observations in advance of some of the first mathematicians in Europe [2; p. 38], the natural philosopher capable of carrying out refined experiments on the quantification of latent heat under different pressure-conditions,[3] the practical chemist who investigated new methods of making alkali from common salt, and invented a new method of testing the degrees of alkalinity, the philosopher who very nearly anticipated Cavendish and Lavoisier on the composition of water[4] – all these are combined in the one man, James Watt, who deservedly earned many international honours in science in Russia, Holland and France, besides his election to the Royal Societies of London and Edinburgh. His invention of the separate condenser must be interpreted against this background, and must be seen as part of the activities of a man who was insatiably curious and devoted to the study of natural philosophy.[5]

[1] James Watt to James Watt junior, 1 May 1785. Birmingham Reference Library (hereafter abbreviated to B.R.L.). See E. Robinson, 'Training Captains of Industry: the Education of Matthew Robinson Boulton [1770–1842] and the younger James Watt [1769–1848]', *Annals of Science*, Vol. X, no. 4, Dec. 1954, pp. 301–13, reprinted in A. E. Musson and Eric Robinson, *Science and Technology in the Industrial Revolution* (1969), pp. 200–15.
[2] Collection of Colonel P. M. Thomas, M.A., D.S.O.
[3] Eric Robinson and Douglas McKie, *Partners in Science: James Watt and Joseph Black* (1969).
[4] James Watt to Joseph Black, 21 April 1783, Doldowlod (hereafter abbreviated to Dol.).
[5] On the general importance of the links between science and technology in the Industrial Revolution, including many references to Watt, see Musson and Robinson, *op. cit., passim.*

The earliest account of Watt's experiments on steam-engines is in Watt's own words in a notebook mainly devoted to the quantification of latent heat [3]. From this notebook it appears that Watt was first interested in the Savery engine but that his attention was diverted to the Newcomen engine by the accident of his being required to repair a model, made by Jonathan Sisson, of a Newcomen engine [Plate 3], for Professor Anderson's class at Glasgow University. This model caused Watt to undertake the first of many experiments upon steam. In these experiments he measured the time taken to evaporate a known quantity of water, devised a new boiler and tested it, measured the expansion of water when turned to steam, discovered the quantity of coals necessary to produce a certain quantity of steam and, as our extract shows, discovered that water in the form of steam could contain more heat than it did when water. This was only the first of many series of experiments upon steam, the importance of which was generously acknowledged by Dr. Joseph Black in his lectures at Glasgow and later at Edinburgh.[1] Among his early fellow-experimenters were John Robison, William Irvine and James Lind, all of them in their different ways well-known figures of the English scientific world in the eighteenth century.[2] Watt's own experiments on the heat of boiling water *in vacuo* (1773–74), on the proportional expansion of steam at different heats, and on specific and latent heats were, for their period, quite sophisticated and undertaken with great care. Some of the apparatus for later experiments on the latent heat of water distilled *in vacuo* is still preserved in the Watt workshop at the Science Museum, South Kensington. Though Eric Robinson and Douglas McKie have shown that these experiments were not faultless, and that Watt's arithmetic could occasionally be fallacious,[3] it still remains true that the experiments were thoughtful, required laboratory technique, and show that Watt's approach to his problems was that of a scientist. It is true that, as Watt admits, earlier engineers, like James Brindley, had experimented in their engines with materials, like wood, of low conductivity, but probably only Smeaton could begin to approach Watt's methodical application of reason to the solution of engineering problems. When John Robison in 1797 proposed a series of scientific experiments on mechanical problems he thought of carrying these out near Soho, Birmingham, not only because he would there have Watt's engines available but also because he wished to discuss his theories with that great theorist of steam-engineering, his old friend Watt. If Dr. James Hutton wishes to make a geological map of Cornwall he writes to Watt; if Dr. Priestley wishes to have a careful observer of his experiments on gases it is to Watt that he turns; if Berthollet wants to know of the practical developments in chlorine bleaching he consults the man to whom he first explained the properties of the gas, James Watt; and the story of Watt's close correspondence with the scientific men of his age could be duplicated many times over from Watt's papers.

What is perhaps more surprising is that Watt did not consider himself to be a superb practical engineer. He states, in his 'Answers to the Objections made to his Specification' [43; p. 195] that:

'He is not so presumptuous as to think that there were not and are not Numbers of Mechanics

[1] This notebook is printed in its entirety in Robinson and McKie, *op. cit.*, pp. 444–89.
[2] See D. McKie and N. H. de V. Heathcote, *The Discovery of Specific and Latent Heats* (1935).
[3] Robinson and McKie, *op. cit.*, pp. 433–43.

in this Nation, who from the same or even fewer, Hints would have compleated a better Engine than he did. Mr. Bramah has proved he could, and Mr. Watt is inclined to believe him. . . .'

He frankly confesses in the same document that in his early years,

'He wanted Experience in the construction of large Machines, that he endeavoured to acquire, but experimental Knowledge is slow Growth, and with all his Ingenuity so much boasted to his prejudice, he was concerned in making some very indifferent Common Engines.'

Among these were the engines at Taylor's Pit and the Schoolhouse Yard at Bo'ness, but Watt in collaboration with a Mr. MacKell also constructed other engines in Scotland. The transference of his ideas from model engines to full-scale engines was for Watt a major difficulty to which he constantly alludes, as in a letter to Dr. William Small, dated 28 April 1769:

'I find that I am not the same person I was four years ago when I invented the fire engine and foresaw even before I made a model almost every circumstance that has since occurred. . . . *The necessary experience in great was wanting*; for acquiring it I have mett with many disappointments. I must have sunk under the burthen of them if I had not been supported by the friendship of Doctor Roebuck. . . .'[1]

As an instrument-maker, Watt had no experience of 'engineering in the vulgar manner'[2] and for years after making his invention, and even after taking out his patent, he was busily engaged in acquiring it. He was always conscious of the gap between theory and practice in engineering, but preferred to think of himself as a natural philosopher first and as an engineer second. He pointed out that Savery's and his own inventions were, as late as 1795, the only '*new applications of philosophical principles*' [41] and stated quite frankly in a letter to his son, James, that:

'the *mechanism* of the engine was not invented all at once, but had been in a course of improvement even unto this day. I was ignorant of many things I know now and possibly of many things which were known to others. Nor do I pretend that I ever possessed an eminently inventive genius in *Mechanicks*. my forte seems to have been reflection and judgement in combining and applying things invented by my self and others.' [27; p. 109]

To this he might have added that he was a great acquirer of book-knowledge at all stages of his career. Robison tells us how Watt read the works of Belidor and Desaguliers at an early stage of his experiments on the steam-engine, and how he acquired some German in order to read Leupold's *Theatrum Machinarum* and Italian in order to read still further scientific and technical works [2; p. 25].

The invention of the separate condenser is the key to Watt's mind and the corner-stone of all his later inventions connected with the steam-engine. It arose from the precise recognition of the basic defect in the Newcomen engine, the wastage of steam involved during the heating, cooling and reheating of the cylinder, and the recognition of the problem was reinforced by Watt's experiments on the reduced boiling-point of water in pressures lower than that of the atmosphere. Watt was perfectly correct in claiming that what he had discovered was a method

[1] Dol.
[2] Small to Watt, 19 October 1771, Dol.

of lessening the consumption of fuel in steam-engines based on a philosophical principle. In the 'Answers by Mr. Watt to the Objections made to his Specification' [43; p. 194] he strictly disclaims originality for all discoveries not made by him:

> 'Of all these things Mr. Watt must say "non ea nostra voco" – The things that are his remain to be told.
>
> He found by the Application of the Knowledge which has been mentioned, that the Cause of the great Consumption of Fuel was that the Cylinder being cooled by the injection Water that Vessel *must* condense a large Quantity of Steam when ever it was attempted to be again filled with Steam; That the Vacuum could not approach to Perfection without the Steam was cooled below One Hundred Degrees, and that such Cooling would increase the Evil complained of in a fourfold or greater Ratio because the Penetration of the Heat or Cold into the Cylinder would be as the Squares of the differences of the Heats between that Vessel and the Steam.'

From that point he proceeded with the inevitability of logic to his solution, and though it may be suspected that others might eventually have arrived at the same diagnosis and solution, the fact remains that Watt was the first to do so and made his invention by no hit-or-miss methods but by the application of reason.

While acknowledging Watt's priority in his invention we should not, however, overlook the general climate of interest in the steam-engine which seems to have been reaching fever pitch in the middle and later decades of the eighteenth century. There is an opinion, which seems to be gaining popularity among historians, that history which pays any special attention to great men is bound to be distorting and misleading, that any great man is the product of the society to which he belongs and that we should therefore look first at general social forces operating at a particular period and only secondly at the individuals thrust into prominence by those forces. In the history of technology and science, such an opinion seems to be closely allied with a school of thought which suggests that a technical problem is solved when a series of social forces coincide in such a way as to provide a moment of ripeness for that solution, and that if one man had not solved the problem another would have done so. How far is this true of Watt and his time? How much credit should be given to Watt as a great seminal mind or, conversely, how much of his fame can only be understood by seeing him as a man who benefits from appearing on the scene at a particularly fortunate moment in time? There were, of course, other important engineers. Smeaton, in particular, by various improvements doubled the efficiency of the Newcomen engine. Dr. Dickinson has described how Smeaton experimented on a model engine, coming to the conclusion that the cylinder should be small in diameter in relation to its length; that he also improved the boring of cylinders; and that he used boilers with internal flues;[1] while even earlier Smiles had stressed the scientific cast of Smeaton's mind.[2] The correspondence from Smeaton which we print in this volume [20] affords similar witness, but it must be acknowledged that Smeaton made no radical departure from the principles of the Newcomen engine.[3] In 1768 Joseph Hateley took

---

[1] H. W. Dickinson, *A Short History of the Steam Engine*, with a new introduction by A. E. Musson (1963), pp. 61–64.

[2] S. Smiles, *Lives of the Engineers* (1874), *Smeaton and Rennie*, pp. 94–95.

[3] J. Farey, *A Treatise on the Steam Engine* (1827), p. 134.

6

out a patent 'for a new Fire Engine and Boiler of a particular sort'. Watt's comment was that:

'The Invention is to keep the Steam Cylinder always *cool* ! ! ! by surrounding it with Water and to keep the boiler very hot, by surrounding it above and below with Flame.'[1]

If this were a completely independent invention, it would suggest that other minds were working on the same lines as Watt, even though there was a basic misunderstanding in Hateley's patent of the principle of condensation and use of steam in the cylinder. The documents which we print below [5, viii, and 13] suggest, however, that this patent only reflects Hateley's misuse of information acquired at Bo'ness while Watt was working there. In 1774–75, when Boulton and Watt were seeking to obtain the Act of Parliament for extending their patent, and when the opposition to their intention was being led in the Commons by Edmund Burke and Henry Seymour Conway, two further challengers appeared – William Blakey and Humphry Gainsborough. Blakey's engine was a modification of the Savery engine[2] and therefore constructed on entirely different principles from Watt's, as Boulton and Watt pointed out in a printed leaflet [14]. Blakey withdrew his objection and was no further problem. After the Act had been obtained, Gainsborough petitioned in 1776 for a patent for a steam-engine but Watt entered a *caveat* against it and the patent was refused so that we do not know the details of Gainsborough's invention. Boulton and Watt made unsuccessful efforts to meet Gainsborough and discuss the matter with him. Gainsborough was extremely annoyed and wrote what is described as his 'anathema' letter to Boulton in 1776. Some impression of Boulton's own lack of principle in such matters may be judged from a statement by him in a letter to Watt that 'I wish you could learn the plan of Gainsborows piston as by his own account tis absolutely perfect for water'.[3] Whatever the effectiveness or otherwise of Gainsborough's invention, it seems to have been original to him and not assisted by previous knowledge of Watt's discoveries;[4] but for the calumnies later propagated by Philip Thicknesse, 'Dr. Viper', a man who later crossed swords with Boulton about Boulton's status as a gentleman,[5] and disseminated by G. W. Fulcher in his *Life of Thomas Gainsborough* (1865), that Watt had learnt of Gainsborough's principles through an unnamed third party, there seems to be no warranty at all.[6] No wonder that Boulton talked of 'Tubal-Cains, or Watts, or Dr. Fausts, or Gainsboroughs, arising with serpents like Moses', that devour all others'.[7]

As the century progressed, so the number of steam-engine patents taken out increased: in 1779 Matthew Wasbrough's 'new invented Machine or Piece of Mechanism' to produce rotative motion from steam-engines; 1780, James Pickard's 'Method of applying Steam

[1] 'Boulton versus Bull. Specifications of Steam Engines', Dol. See also *Patents for Inventions. Abridgements of Specifications relating to the Steam Engine* (1871), Part 1. A.D. 1618–1859, Vol. I, pp. 56–57.
[2] J. J. Bootsgezel, 'William Blakey – A Rival to Newcomen', *Newcomen Soc. Trans.*, Vol. XVI (1935–36), pp. 97–110.
[3] May 1775, Dol.
[4] 'Notes made on reading the letters and papers relating to Mr. Gainsborough's patent in Embrio – which proved an *Abortion*', in Watt's hand, Dol.
[5] Matthew Boulton to Philip Thickness, ?1781, Assay Office Library, Birmingham (hereafter abbreviated to A.O.L.B.).
[6] J. P. Muirhead, *The Life of James Watt* (1858), pp. 101–06.
[7] Boulton to Watt, ?1776, quoted Muirhead, *op. cit.*, p. 105.

Engines to the turning of Wheels'; 1781, Jonathan Hornblower's steam engine; 1784, Robert Cameron's 'certain Methods of making and constructing a steam engine'; 1787, William Symington's 'Steam Engine on Principles entirely new'; 1790, Adam Heslop's 'Engine for lessening the consumption of Steam and Fuel in Fire or Steam Engines'; 1790, John Westaway Rowe's 'Engine to be worked by the power of Steam on an entire new Construction'; and 1792, Francis Thompson's 'Steam Engine on a new Construction'. In collecting copies of patents as evidence in their various cases, Boulton and Watt actually listed thirty-four patents for steam engines, besides Watt's, between 1698 and 1793, but stated in 1795 that they believed that there were nearly a dozen more, 'chiefly of late, the titles of which are imperfectly known'.[1]

It should not be overlooked, however, that many of these patents were taken out by men who had access, in one way or another, to Watt's inventions: Joseph Hateley worked for Dr. Roebuck at Carron, Wasbrough and Pickard had premises at Snow-hill, Birmingham, and obtained information from one of Boulton and Watt's workmen, Richard Cartwright; Jonathan Hornblower and Robert Cameron had been engine-erectors for Boulton and Watt; while the other patents listed above, as well as several others, were all claimed by Boulton and Watt to be piracies of their engine. Symington's engine, as described by Dr. James Hutton,[2] used two cylinders in the second of which condensation took place so that it infringed the patent for the separate condenser, and the same evasion was practised by Hornblower and by Francis Thompson. John Westaway Rowe is mentioned in a letter from Samuel Wyatt, brother of James Wyatt and himself an architect, to Matthew Boulton in 1790,[3] from which it appears that Rowe had entered the employ of John Wilkinson at Bersham, the man who above all others acted traitorously towards Boulton and Watt, in that, while he enjoyed a monopoly of the production of cylinders for their steam-engines, he yet used their patent without licence in some of his own engines and encouraged others to infringe also. Boulton and Watt's engines were so famous, nationally and internationally, that an engineer who had no experience of them could hardly claim to be adequately informed, let alone a leading member of his profession. Evidence of the international fame of the Boulton and Watt engines is provided not only by the enquiries of foreign customers, such as M. Jary of Nantes in 1779 and the Perier Brothers of Paris in 1780, but also by the numerous continental industrial spies, such as Baron von Stein, Baron von Reden of Clausthal, the Comte de Bétancourt, Dr. Joseph von Baader and others who from the 1780s onwards were a constant menace to Boulton and Watt.[4] At home in Britain the number of people with alternative suggestions for steam-power and of infringers also grew steadily. Boulton refers to the schemes of the 'Mr. Hateleys, Mr. Wasbroughs, Mr. Matthews, Mr. Pintos, Mr. Jones, the

---

[1] Boulton and Watt *vs.* Bull, 'Specifications of Steam Engines', Dol.

[2] James Hutton to James Watt, 23 September 1788, Dol.

[3] Samuel Wyatt to Matthew Boulton, 1 March 1790, A.O.L.B.

[4] See W. O. Henderson, *Britain and Industrial Europe, 1750–1870* (Leicester, 1954), p. 150, n. 47; J. Payen, 'Documents relatifs à l'introduction en France de la machine à vapeur de Watt', *Revue d'histoire des Sciences*, Vol. XVIII (1965), pp. 309–14, and 'Bétancourt et l'introduction en France de la machine à vapeur à double effet (1789)', *Revue d'histoire des Sciences*, Vol. XX (1967), pp. 187–98.

Spaniards and the Truro mans'.[1] 'The Spaniard's' engine was being puffed by Adam Walker, the itinerant lecturer in science, in December 1781,[2] but his claims to originality, like those of other patentees, become very suspect once we examine the provenance of their ideas. Only one patent quoted against Boulton and Watt in their legal cases seems to have taken them by surprise, and that was the patent taken out in 1759 by the Rev. Henry Wood for the use of an air-pump to extract hot air or condensed vapour from the cylinder, but not from a separate condenser. The engine seems never to have been brought into an actual working state, but the patent helps to substantiate the suggestion that others were experimenting along similar lines to Watt, and independently of him.

In one letter Watt himself acknowledges, in a manner unusual for him, that the spirit of enterprise flourishing in the age, and not only the thieving propensities of mankind, may have had a part to play in thrusting up rivals [25]:

'I do not think we are safe a day to an end in this enterprizing age. Ones thoughts seem to be stolen before one speaks them. It looks as if Nature had taken up an adversion to Monopolies and put the same things in several heads at once to prevent them.'

This was what Robison pointed out in his evidence in 1796 [2; p. 25] when he said that Watt had:

'disclosed so much of what he had been doing, that had it been in London or Birmingham, I am confident that two or three patents would have been expeded for bits of his Method by bustling Tradesmen before he thought himself intitled to solicit such a thing.'

As we shall see, this was certainly Watt's experience once he left Scotland.

The turning-point in Watt's career was his departure from Scotland for Birmingham, and the commencement of his partnership with Matthew Boulton. Even Dr. James Hutton, who was anxious as any man to retain Watt's services in his own country, was obliged to admit that an alliance with Boulton offered a greater certainty of a successful career in engineering for Watt than anything that Watt's friends were able to contrive for him at home.[3] Birmingham might be 'that place of buttons and japann'd crokery', but it was preferable to the hard hills of desolation in Scotland.

On 8 July 1768 Robison wrote to Watt about a visit that the former had made to Birmingham:

'I was much struck with many appearances of ingenuity in Boulton's works. It is partly at his desire that I now write to you. In hopes that you had vigorously pursued the improvements of your fire Engine, and was getting your property secured, he has finished a building for one of your reciprocating Engines, and has waited delayd [*sic*] the erection of it on your account because he says that he would be sorry to put the world in possession of an improvement of which you are the Author. He is however in very great want of it, and [may] be obliged by necessity to finish it. . . .'[4]

[1] Matthew Boulton to James Watt, 1 July 1780, B.R.L.
[2] Gilbert Hamilton to James Watt, 15 December 1781, Dol.
[3] Dr. James Hutton to James Watt, February 1775, Dol.
[4] John Robison to James Watt, 8 July 1768, Parcel A, B.R.L.

The contribution made by Boulton to the success of Watt's engines can hardly be over-estimated, though the part played not only by Dr. Roebuck but also by Dr. Black should not be overlooked. Watt's debt to Black was not repaid until 1781 and no one could have been a better friend or a more patient creditor than Black. But what Watt required was not only a sympathetic friend and a financial backer but also someone to compensate for his own lack of experience in the metallurgical industry. In all these respects Boulton was the perfect partner. His letter to Watt rejecting a licence for a few counties but offering to make engines for the whole world is one of the greatest documents of the Industrial Revolution [9]. It demonstrates the breadth of Boulton's vision, which was fully justified in the event, the incisive judgement which allowed him to generalize from his own needs for power to the probability of a world demand, and the perception of the totally different character of Watt's inventiveness from that of the ordinary empirical engineers of the day. Boulton's far-sightedness was demonstrated again in his insistence that Watt should not be satisfied with ordinary pumping engines but should invent an engine producing rotative motion directly [22]. So much is common knowledge, but the relationship between Boulton and Watt, the classic entrepreneur-engineer team of the Industrial Revolution, is more complex than such generalizations suggest, and has been obscured not only by the tendency of historians to repeat each other's judgements, but also by Watt's memoir of Boulton which described a more continuously peaceful relationship than was possible even for a David and Jonathan.

First of all we should not allow ourselves to be deceived by our own terminology. Though Boulton and Watt formed an alliance of entrepreneur with engineer, this does not mean that Boulton was devoid of engineering experience or Watt totally without financial acumen. A volume of mechanical copies of letters at Doldowlod headed 'Watt's private letters to Matthew Boulton and James Keir' shows how much work was done on the engine by Boulton, in Watt's absence, on the improvement of such things as the fit of the piston in the cylinder, the pressure of steam from the boiler, and even on improvements of the condenser. Such a letter as the following is not uncharacteristic:

> 'Our copper bottom hath plagued us very much by steam leaks and therefore I have had one Cast (with its conducting pipe) all in one piece, since which the Engine doth not take more than 10 feet of steam and I hope tomorrow to reduce that quantity as we have just received the new piston which shall be put in it at work tomorrow.'[1]

Boulton and Watt joined forces largely because they were both natural philosophers and both practical men, sharing interests in chemistry and mineralogy as well as in steam-engines. It should be remembered, however, that though Watt was understandably obstin-ate about accepting engineering suggestions from others, including Boulton, he could on occasion be worn down, as when William Murdock and James Watt junior finally per-suaded him to prefer the slide-valve to the puppet-valve [4; p. 44], or when Boulton con-vinced him that he should proceed with rotative motion.

The converse of this is that Watt occasionally took decisive action in financial matters. Rightly or wrongly he would not allow Boulton to mortgage their engine-agreements fur-

---

[1] Matthew Boulton to James Watt, 10 July 1776, A.O.L.B.

ther than they already were by June 1780 and by so doing obliged Boulton to disentangle the affairs of Boulton and Fothergill from those of Boulton and Watt. At the same time he insisted upon approving all expenditure made on the account of Boulton and Watt and even suggested that Boulton should sell out part of his interest in the partnership to a new moneyed partner.[1] Since, as J. E. Cule has shown, it was Boulton and Watt that was making money and Boulton and Fothergill that was losing it, Watt had some justification for his intransigence.[2] Though he did not possess Boulton's essential flair for obtaining credit from merchants and bankers, Watt had one advantage over Boulton as a businessman. He was single-minded, whereas he had rightly regarded Boulton from the very first as something of a projector, inclined to pursue one scheme after another. Watt was interested in selling the steam engine and thus obtaining a substantial and secure income. Unlike Boulton he was essentially a modest man and though, in his old age, honours flowed in upon him, he seems to have been sublimely indifferent to status and prestige. From the first, Watt tells us in his 'Answers . . . to the Objections made to his Specification' [43; p. 195], 'his mind run upon making Engines *cheap*, as well as *good*', and he had no desire to rival Boulton as a Maecenas-like figure, dispensing patronage and accumulating honour.

Watt's material fortune depended not only upon the efficacy of his partnership with Boulton, but also upon the state of the patent law. Dr. Dickinson believes that Boulton and his friend Dr. William Small wrongly advised Watt over his first patent and that consequently the engineer made a mistake in taking out a patent for a principle instead of for an application of a principle.[3] The truth is much more complicated, and Watt spent another thirty years after 1769 trying to find equitable solutions to the problems thrown up by the existing state of the patent law. As we have seen, what Watt discovered was a method for lessening the consumption of fuel in steam engines, and this method consisted in condensing the steam in a vessel separated from, but communicating with, the cylinder. The method arose from the contemplation of scientific principles and was not primarily a mechanical improvement, though the perfection of the method on a larger scale in steam engines proper and not merely in models did create mechanical problems, most of which Watt claimed to have foreseen. It is clear that Watt had intended at first to submit an application for a patent for a particular engine, and his drafts [6] show him describing different engines in some detail and proposing to accompany the description with drawings. Boulton and Small dissuaded him from his first ideas by proposing instead a patent of a much more general nature [7] related to the *method* rather than to the *engine or engines*:

'Mr. Boulton and I have considered your paper, and think you should neither give drawings nor descriptions of any particular machinery, (if any such omissions would be allowed at the office) but specify in the clearest manner you can, that you have discovered some principles, and thought of new applications of others. . . .'

The intentions behind this procedure are quite plain. If Watt were to patent a certain engine,

[1] James Watt to Matthew Boulton, 21 June 1781, A.O.L.B.
[2] J. E. Cule, 'Finance and Industry in the Eighteenth Century: The Firm of Boulton and Watt', *Economic History*, June 1940.
[3] H. W. Dickinson, *James Watt, Craftsman and Engineer*, p. 52.

or engines, with drawings, etc., it would be easy to circumvent the patent merely by making some small adaptation of the mechanism in order to produce a superficial difference of design while continuing to make use of Watt's principle, and this was what most of the pirates in fact proceeded to do. At the same time Boulton and Small were not crystal-clear about their intentions since they spoke not only of new 'principles' but also of 'new applications of others' and gave in the rest of the draft unnecessary detail about the piston-packing, the lubricants employed and other matters irrelevant to the principle. In addition, when the 1769 patent was extended by Act of Parliament in 1775, Watt submitted detailed drawings and descriptions. Moreover, the Act mentioned certain 'engines' which Boulton and Watt had taken some time to perfect and which had involved them in considerable expense. Their legal opponents argued that Watt could not have a patent both for a principle and for certain specified engines, and that the Act of 1775 was not simply an extension of the patent granted in 1769, but a new instrument conferring upon Watt rights in particular engines. Thus whatever the intention of the patentees in 1769, that intention was at least made to appear confused by the Act passed in 1775, which seemed to do precisely what Boulton, Small and Watt had tried to avoid six years earlier.

A further difficulty was that the intention of granting a patent to an inventor was not only to ensure a monopoly for the inventor, but also to release his invention to the general public after the term of the patent had elapsed. For this reason the specification was required by the courts to be sufficient for the invention to be understood and used by others, though Watt argued that it was originally only a method of distinguishing one patent from another and that there was no warranty for it in statute law. Who the others should be who were to be capable of understanding the specification was not in any event defined. Thus Small could write to Watt [7]:

> 'I am certain that from such a specification as I have written any skilful mechanic may make your engines, altho it wants corrections, and you are certainly not obliged to teach any block-head in the nation to construct masterly engines.'

As we show below [39–45], this became a major point of difference between Boulton and Watt and their opponents at law: Boulton and Watt claiming that their specification was sufficient to meet the requirements of the law and their opponents denying it. Ultimately, after some seven years of litigation the issue was determined, legally, in favour of Boulton and Watt, but certain moral issues remain unresolved by the legal decision.

How far were Boulton and Watt concerned to draw up a sufficient specification and how far were they concerned to reveal as little as possible in the specification? Perhaps there is a hint of an answer in a letter from Watt to Small in August 1773:

> 'I have some thoughts of writing a book, the 'Elements of the Theory of Steam-engines', in which, however, I shall only give the enunciation of the perfect engine. This book might do me and the scheme good, and would still leave the world in the dark as to the true construction of the engine.'[1]

It is true that this statement does not refer to specifications, but it is written after the 1769

---

[1] James Watt to Dr. William Small, 17 August 1773, Dol.

specification had been enrolled and breathes that secretive atmosphere which is such a characteristic part of eighteenth-century business-life. If the patent law of the eighteenth century had been a stronger shield against infringements, there might have been less reluctance on the part of inventors to reveal all in their specifications. Despite the fact, for example, that it was possible for anyone to enter a *caveat* against another's patent if it was felt that the patent was invalid by reason of being an infringement or because it sought to gain a monopoly of a method or machine already commonly known, and despite the fact that Boulton and Watt successfully entered such a *caveat* against Gainsborough's patent, the Attorney General granted some very remarkable patents. Many of the steam-engine patents granted infringed Watt's patent, but beyond this, Pickard's and Wasbrough's patents were, in Watt's opinion, totally invalid for two reasons: first because they were based on information stolen from him, and secondly because they sought to patent the crank, a device known to mankind since the days of ancient Egypt, or, as Watt wrote to Boulton [21, vii]:

> 'being thoroughly convinced in my own mind that they invented no part of it. they got the hint of the Crank from Ned Ruston who had it from a lathe and the weight they got from us via Cartwright.'

Why then did Watt not challenge them? Because he was frightened that if he were to challenge other people's patents he would stir up people to question his own patent. Yet this fear did not arise from any lack of confidence in the legality of his own patent but from a realistic, and justified, appraisal of the weaknesses of the patent law in the eighteenth century and a suspicion, common to many of us, of lawyers and all their ways. Consequently Boulton and Watt did not challenge Hornblower's patent until Hornblower sought to extend it, in imitation of Watt, by Act of Parliament [35–36]. Watt's partner, Boulton, had extensive Parliamentary influence and had reason to have more confidence in his 'friends' there than he had in the ordinary courts of law.[1] Moreover, Watt was fully informed about Henry Cort's and Richard Arkwright's lack of success in defending their patents at law, and was indeed a witness for Arkwright in 1785 when Arkwright's patent was overthrown. Watt later wrote [48; p. 227]:

> 'even granting that the greater number of Patents shou'd prove useless both to the state and to the inventors, *if one Capital and meritorious invention such as Sir Richard Arkwright's Cotton machine shou'd be brought forth in a Century in consequence of them* it shou'd justify the measure of granting them.'

It is a measure of the spirit of the man that when frustrated and, as he felt, robbed by the patents granted to Pickard and Wasbrough [21], Watt did not turn to the law but to his own inventive fertility to find a way out of his *impasse*. He thus devised five methods, including the sun-and-planet gear, for obtaining rotative motion [23]. A number of inventions, including the governor, the earliest instance of a feed-back mechanism, were not, however, secured by patent at all [33].

[1] Even in Parliament, however, the issue was not a foregone conclusion: 'I suppose we shall be out Voted in the Commons but let us try the honour of the Lords.' Boulton to Watt, 26 March 1792, B.R.L.

Watt's decision not to take out patents for all his inventions, or, in other cases such as his improved furnace [30], to delay taking out the patent, may be explained by the difficulty, experienced not only by Watt but also by Henry Cort and Sir Richard Arkwright, of sustaining patents in which certain processes were combined in a new way or in which a new application was made of an old device or principle. Cort's and Arkwright's patents were overthrown because it was not clearly accepted by the courts that a new combination of known processes could be protected by patent. Watt's governor, similarly, was an adaptation of a device used in wind-driven flour mills to maintain a constant and regular drive despite fluctuations in wind-speeds, and perhaps for this reason was not patented.

In 1793 Watt was asked to comment on a new patents bill.[1] He did so at very considerable length, and of three different drafts known to us, we have printed the lengthiest which seems to be the most complete and a fair copy [48]. In this document he listed the persons who, in his opinion, were entitled to patents and includes among them:

'Those who combine together old Instruments or machines so as to produce new effects, or to make them more extensively useful to the publick –
Those who apply old machines or instruments to new uses – Provided such new uses be essentially different from the common uses of the said machines or Instruments.'

It is in this same document, one of the most interesting in the history of the patent law, that he pointed out that specifications did not appear before 1714, were not required by statute law, and were used in the first place merely to distinguish one patent from another. Watt suggested that for the greater security of the inventor the time during which a patent might be challenged should be limited; that there should be a body of technically and scientifically qualified Commissioners to adjudicate on certain matters relating to patents; that an inventor should be allowed to refine on his original patent as a consequence of his increasing experience of it and add additional matter so long as these changes did not constitute a fundamental alteration in the nature of the original patent. It was a matter for debate in Boulton and Watt's legal cases how far a patentee could be expected to have completed all the experimental work and to have produced a finished machine or method at the time of the enrolment of the specification. The witnesses in the Hornblower Case testify that they could not erect a Watt engine without having completed many experiments, while Watt himself argued that it could not be expected that at the moment when a patent was taken out all the refinements of the invention would be in the inventor's mind. Problems of this kind Watt attempted to solve in his proposals for new legislation.

The actual conduct of the various legal actions and of the political manoeuvres in Parliament in 1775 and 1792 is too complicated a matter to be dealt with fully here.[2] It cannot be fully appreciated, however, without some reference to the kind of society in which these conflicts took place. It was a society dominated by patronage and it was certainly no small factor in Watt's eventual success that he was allied to a man who was a master of social and political manoeuvre. Boulton's lines of influence, observable in his vast range of corres-

[1] James Watt to J. Anstruther, 20 May 1793, Dol.
[2] For 1775, see E. Robinson, 'Matthew Boulton and the Art of Parliamentary Lobbying', *Historical Journal*, Vol. VII, no. 2, 1964.

pondence, were so extensive that it is almost true to say of eighteenth-century society from 1760 onwards that if Boulton did not know you or did not know of you that you were probably not worth knowing. The consequence was that in the various political conflicts that arose about Watt's Fire Engine Act of 1775 and Hornblower's bill of 1792 Boulton could summon scores of friends in both Houses of Parliament to his support. It is also true that though Watt frequently complained of Boulton's tendency to cultivate the nobility and influential persons generally, the tactics of the legal cases were to a large extent directed in a conscious way to the demonstration of the social superiority of Boulton and Watt over their rivals. As witnesses they summoned some of the leading scientists of the age including Herschel, the astronomer, Professor Robison, and J. A. de Luc, and Bramah was quite right to question the validity of their evidence on the technicalities of engine-construction [45]. No expense was spared in taking affidavits, paying the expenses of witnesses, publishing pamphlets and advertisements, having models constructed to demonstrate the mechanical principles involved, or in briefing the best available counsel. Though Watt's cause was unquestionably a just one, there was no gap left in the defences of the patent, and Watt clearly spent a great deal of time in drawing up relevant accounts of his invention to assist his lawyers, while Boulton kept the artisan witnesses amused 'by shewing them sights and Engines in order to keep them out of the Enemys Camp'.[1] From the first, too, Boulton was anxious to enlist the support of the *cognoscenti:*

> 'Perhaps it might not be impolitical if you were to publish a paper in the Philosophical Transactions chiefly Elementary intimating that We have a variety of Engines invented very different in their construction some where the piston is pressed upwards and without the great Beam others where there is a constant Vacuum under the piston . . . and that you have annexed a drawing of one (which is *Erected* at Bedworth and that it *doth so and so* with such a quantity of Coals). You may compliment the York building Engine and say tis the best you have ever examined. . . . The Curve of Boiling points under different presures [*sic*] will do you honor if you think it prudent to publish it. . . .'[2]

Here is a master tactician at work.

Yet tactics alone would have been of doubtful efficacy. Watt's great reputation was based on solid achievement in engineering, and that achievement moreover had an inner momentum, a logical consistency, even a style which demonstrates the impact of a strong controlling intellect. To begin at the beginning, with the invention of the separate condenser: that was an invention which could be applied with effect to the Newcomen atmospheric engine, and by other engineers was so applied. There still remained a considerable heat loss, however, since the cylinder was exposed to the atmosphere. Brindley, and later Watt, tried to overcome this heat loss by using cylinders of wood and therefore of low conductivity, but the drawbacks to this material are obvious. There was also a heat loss due to the normal practice of using a layer of water above the piston in the open-topped cylinder so as to secure an air-tight fit. Even if hot water was employed for this purpose, as it sometimes was, to reduce the heat loss, there was still the problem of the water seeping into the cylinder and

[1] Matthew Boulton to James Watt, 26 March 1792, B.R.L.
[2] Matthew Boulton to James Watt, 10 July 1776, A.O.L.B.

being converted into steam, making complete condensation difficult. Watt decided to en-close the whole of the cylinder in a steam-case, which meant, among other things, that he was obliged to seal off the open top of the cylinder and do away with the water floating above the piston. The piston now had to move through a stuffing-box and it was also essen-tial for it to have as tight a fit as possible in the cylinder – hence his proposals to use oils or fats or even mercury. What is more, the piston could no longer be depressed in the cylinder through the action of the atmosphere now that the cylinder was sealed off and so Watt used steam-pressure to do the work of the atmospheric pressure. All kinds of experiments were undertaken in order to improve the piston-packing, and it is true that experiments of this sort were entirely empirical. They arose, however, from a problem which followed as a strict logical consequence of the need to keep the cylinder constantly hot.

Watt had now produced an engine which was much more economical in fuel consump-tion than Newcomen's, firstly by condensing in a separate condenser kept permanently cool, and secondly by keeping the cylinder permanently hot. But his engine was still only single-acting, *i.e.* working power was applied only on the down-stroke of the piston. He had substituted the power of steam for atmospheric pressure to force down the piston, after a vacuum had been produced below it by condensation, but the piston was then raised, after readmission of steam into the cylinder, by the weight of the pump-rods or counterweight at the other end of the beam, as in the Newcomen engine. Moreover, all the early Boulton and Watt engines, again like Newcomen's, were simply reciprocating pumping engines (pumping water or air). Watt's idea of a 'steam-wheel' to produce 'circular' or rotative motion – part of his 1769 patent and specification [6 and 8] – eventually proved abortive, after experiments in the early 'seventies [15 and 16], though Watt still did not abandon the idea. It was not until later in that decade that he turned his mind seriously to the conversion of the tradi-tional beam engine's reciprocating action into rotative motion. Various people had made earlier attempts in this direction, using cumbersome arrangements of ratchets, pinions, etc., but not with any great success, as Watt noted in the late 1760s [6], and the common way of using steam engines for driving machinery was by the indirect method of raising water to turn a water-wheel. Another obvious way of harnessing steam power directly, however, was by using the ancient device of the crank. This idea had occurred to Watt at least as early as 1771,[1] and by 1779 he was experimenting with a double-cylindered engine acting on two cranks, together with a flywheel and counterweight [28], but, as we have previously men-tioned, this arrangement was apparently stolen from him and patented by Pickard in 1780. This infuriated Watt and he seriously considered taking legal action to quash the patent, but instead he devised various other methods of producing rotative motion, including his famous 'sun-and-planet' device, patented in 1781 and specified in 1782 [23].[2]

At the same time Watt also introduced various supplementary improvements. One of these was to make the engine 'double-acting', by applying steam alternately below as well as above the piston, to produce a power stroke in both directions. We know, from a drawing

[1] Muirhead, *op. cit.*, Vol. III, p. 38.
[2] He also included another version of his 'steam-wheel', but this never came into use and all steam engines, both rotative and pumping, continued to be beam engines.

presented to Parliament in 1774–1775, that Watt had already by that time produced plans for a double-acting engine and may have thought of it even earlier. He delayed doing more about it, however, until 1781 when Boulton proposed it as a means of evading Pickard's crank patent.[1] Watt then took out another patent in 1782 in which he included a double-acting engine [26]. The first such engines to be built were, in fact, rotatives, though the same principle was soon applied to pumping engines.

Another innovation, included in the same patent specification, was that of working steam expansively, about which Watt had been writing to Small as early as 1769.[2] The recognition of the elasticity of steam and of its expansive power is integral to the very earliest of Watt's experiments, while the idea of using that principle in order to economize the use of steam in the engine, by cutting it off before the end of the piston stroke, is a further demonstration of the logical cohesion of Watt's inventions. In actual practice, however, the experiments in 1777 on the engine called 'Beelzebub' were far from satisfactory, so that by 1779 Watt's interest in the idea of expansive working began to flag. It was revived, however, by the appearance in 1781 of Hornblower's compound engine in which the steam was used successively in two cylinders. Hence Watt included the application of the expansive principle in his 1782 patent: firstly by using a cut-off valve, etc., in a single-cylinder engine, and secondly by linking together two engines in a double or compound engine, whereby the steam from the first cylinder was applied expansively in the second [26]. By 1784, however, Boulton and Watt had come to the conclusion that there was little to be gained from this principle – largely because of the ignorance of engine-men [4; p. 43] – and it was in fact abandoned from then onwards. The reader's attention, however, should be drawn to the diagram accompanying the 1782 patent specification [See Colour Plates]: this is based on Boyle's theory of the expansion of gases and is a good example of the scientific way in which Watt's mind was accustomed to work. Some of his ideas, however, were too sophisticated for the practical engineering experience of the time and had to be abandoned 'solely for the sake of keeping the Engines as simple as possible' [4; p. 44].

Nevertheless, Watt's patents of 1781–82 mark the beginning of a new and revolutionary era in the development of steam power, since his engines, now made increasingly powerful by their double-action, could be applied to provide rotative motion of all kinds, from winding coal to working spinning machines, in addition to the continued extension of their use as pumping engines. A necessary adjunct of the double-acting engines, however, whether rotative or pumping, was some means of connecting the piston-rod rigidly to the beam, in place of the flexible chain, so that the engine could exert its 'push' when the piston was forced upwards in the cylinder, in addition to its 'pull' on the down-stroke.[3] In his 1782 patent specification, therefore, Watt also included the employment of a toothed sector-and-

---

[1] Boulton to Watt, 20 April 1781, B.R.L., and Watt to Boulton, 16 Jan. 1782, in Muirhead, *op. cit.*, Vol. II, pp. 137–138.

[2] H. W. Dickinson and R. Jenkins, *James Watt and the Steam Engine* (Oxford, 1927), p. 120.

[3] In both pumping and rotative engines, of course, the beam was still used, and the connection between piston-rod and beam was the same; the difference was in the connection at the other end of the beam – in the former case to a reciprocating pumping mechanism, in the latter to a rotative arrangement such as the sun-and-planet device or a crank.

rack [26; p. 106], but this was a clumsy expedient and sometimes the teeth were broken out of the rack.[1] Watt therefore turned his mind to achieving a better solution and soon produced the idea of 'parallel motion', a mechanical invention of which he himself was particularly proud. He was apparently experimenting with this idea as early as 1781 [28], but did not patent it until 1784 [29; p. 114], at first as a three-bar motion, which was then almost immediately developed into parallel motion proper, in the form of a jointed parallelogram. Dickinson and Jenkins seem to cast doubt on whether parallel motion in its final form was Watt's own invention, but there seems to be no reason for scrupulosity on this particular point, since it is entirely in line with Watt's interest in the pantograph, later developed in his sculpturing machine.

Here again, one sees the difficulty of trying to decide whether or not the application of an old principle to a new use is to be regarded as a new invention. Similarly, the Hornblowers and Watt were involved in dispute over the prior use of the blowing-valve for clearing the engine of air and water when starting it. It was claimed by Jabez Hornblower in Gregory's *Mechanics* (1809) that the blowing-valve was applied first by Jonathan Hornblower, the father of Jabez, to an engine at Ting Tang mine in 1778, but Murdock recalled it being used on the Soho engine in 1777 [28; p. 110]. Watt asserted that [27; p. 109]:

> 'I cannot recall whether the blowing valve was used first at Tingtang Engine or not. It was however no invention of Jabez Hornblower being an essential part of Newcomens Engines and perfectly well known to me.'

There was a difference, however, between the 'snifting' valve of the Newcomen engine, used after each stroke in blowing the air out of the cylinder, and the use of the blowing valve in clearing Watt's engine when starting it. The counter claims expressed in this controversy provide a further illustration of the difficulty of establishing priority of invention where what is being done consists in adapting a piece of mechanism from one use to another related one.

What cannot possibly be illustrated from a short account such as this, but is very adequately described in the magnificent book by Dickinson and Jenkins already referred to, is the constant experimentation upon every part of the Boulton and Watt engines from the valves and the working-gear to the mechanisms for rotative motion, and from the earliest variations in the form of the condenser to the latest improvements in boilers. Watt was not responsible for all the improvements that took place. In the last years of the patent it is clear that William Murdock and John Southern played a very important part in introducing improvements, but the amazing fertility of Watt's own mind until his time was almost wholly occupied with litigation is the thing that constantly forces itself upon our attention. Besides his work on basic design, we should not overlook his concern with the testing of materials,[2] as evidenced by his 'Blotting and Calculation Book 1782 to 1783' in the Birmingham Reference Library, where he recorded experiments on the breaking-points of cast iron, wrought iron, oak and deal. The double-acting engine, for example, subjected the beams of

[1] Watt's *Journal* (1783), 31 March 1783, Dol.
[2] Dickinson, *James Watt, Craftsman and Engineer*, pp. 141–44.

steam engines to new strains and Watt was obliged to redesign his beams in order to overcome this new problem. It is necessary also to defend Watt from the accusations sometimes brought against him as being responsible for the air-pollution caused by the increased use of steam-power. The letter from Dr. Thomas Percival [31] shows how early the evil was recognized by the physicians of the eighteenth century and that one of them at least turned to Watt for the solution of the problem. Watt's proposals were patented in 1785 [30] and he wrote to his friend, De Luc, on 10 September 1785:

> 'I have some hopes of being able to get quit of the abominable smoke which attends fire-engines.'[1]

Watt seems to have been first moved to undertake his experiments by the damage being done to Boulton's garden at Soho, a garden which, by the way, no less a gardener than Humphry Repton had given advice upon. Had the rest of Manchester been as aware of Watt's improvements as his fellow philosophers, such as Percival and James Watt junior's friend Dr. Ferriar, some of the city's blackened appearance might have been avoided altogether. A man who lost two of his children from consumption and spent a considerable time in devising machines to assist the medical profession was not likely to be the sort of man to pursue personal profit at the expense of his social conscience.

By 1800 Watt had enjoyed a patent for his steam engine for thirty-one years. The effect of this long term has been vigorously debated from the eighteenth century down to the present day. Dr. Dickinson came to this conclusion:

> 'We are of the opinion that Watt's master patent for the separate condenser, with the extension granted to him by Parliament, amounting in all to thirty-one years, was unduly long in the public interest. It had tied down progress to the wheels of Watt's chariot, for it must be remembered that he frowned upon any increase of steam pressure beyond a few pounds above the atmosphere on the grounds of safety. With him enterprise had stopped: on the other hand, we have to remember that but for the protection afforded by the patent, Watt would not have enjoyed the environment in which he was able to work out his equally brilliant inventions, necessary to bring the rotative engine into being, and for this we must always be thankful.'[2]

If Dickinson had known of Watt's attempt to extend his patent for a further period after 1800 [47], he would have been, no doubt, even more stringent in his observations, but it is a matter about which it is particularly difficult to come to a conclusion. Boulton and Watt's costing-out of the financial advantages to the nation of their engine set against their own profits minus the cost of expensive litigation [46] is a sobering document. Watt's invention, after all, has been generally recognized as the key-invention of the Industrial Revolution. Its effects upon the copper-mining industry before the exploitation of the Parys Mine at Anglesey was nothing less than revolutionary; its contribution to the cotton industry was equally fundamental, and without the regularity of motion produced by Watt's rotative engines, it seems unlikely that the industry could possibly have expanded at such a phenomenal rate; collieries did not depend to anything like the same extent as the copper or the cotton industries on Watt's engine, but the value of Watt's engine to the iron industry as a

---

[1] Muirhead, *op. cit.*, Vol. II, p. 204.
[2] Dickinson, *Short History of the Steam Engine*, p. 88.

motive force for blowing-engines, hammers, and rolling mills was reflected in the increased demand for coal; canals, breweries, starch-factories and several other industries also employed Watt's engine, but the nineteenth-century growth of railways had, of course, to wait on the development of the high-pressure engine. Such a contribution to the growth of the economy was previously unparalleled, and when it is recognized that, even after the termination of the patent, the Watt beam-engine continued to be the most popular type of power-unit for almost half a century, the patent does not seem unduly prolonged.

On the other hand, as we have shown elsewhere, the contribution of Boulton and Watt to the Steam Revolution has often been exaggerated, notably by J. Lord in his *Capital and Steam Power* (1923) and by many who have followed him.[1] The very large number of Newcomen and also of modified Savery engines that were used in many industrial processes has never been adequately appreciated: certainly far more of them were built in the eighteenth century than Boulton and Watt erected.[2] Moreover, to them must be added the very considerable number of 'pirate' engines, infringing Watt's patents. Taking all these engines into account, it appears that in Lancashire – the main market for Boulton and Watt engines in the later part of the century – the Soho firm had contributed no more than a third of the total, perhaps less;[3] while in Cornwall they could account for barely half.[4] One must also bear in mind the later development of steam power, which makes these early advances appear puny by comparison: by 1800, when Boulton and Watt's patent monopoly expired, they had erected about 500 engines, with a combined horse-power of about 7,500, out of a total which cannot have been much more than three or four times as great; but by 1850 the overall total of industrial steam power (excluding railway locomotives and ships) is estimated to have been about 500,000 h.p., and by the time of the first Census of Industrial Production in 1907 it had soared to 9,650,000 h.p.[5] The nineteenth century is the Age of Steam, not the eighteenth, at the end of which, in fact, water-power still greatly exceeded that of steam.[6]

It would be easy to conclude that the comparative slowness of the early development of steam power was due to Watt's patent. It has been suggested, for example, that Boulton and Watt's monopoly delayed the development of the high-pressure engine, and no doubt it did stand in the way of such engineers as Hornblower, Trevithick, etc.; Bramah complained that Watt's patent was made deliberately general and vague so as 'to lock up the brains and hands of every inventive genius' [45; p. 207]. But it may also be argued that the inadequacies of boiler-making and of engineering techniques were an equally serious obstacle. There is no doubt that explosions in high-pressure engines remained common far into the nineteenth century:

[1] Musson and Robinson, 'The Early Growth of Steam Power', *Economic History Review*, 2nd ser., Vol. XI (1958–59); see also *Science and Technology*, pp. 393–426.

[2] In addition to our publications on this subject, see J. R. Harris, 'The Employment of Steam Power in the Eighteenth Century', *History*, Vol. LII (1967), pp. 133–48.

[3] Musson and Robinson, *Science and Technology*, p. 426.

[4] J. Rowe, *Cornwall in the Age of the Industrial Revolution* (Liverpool, 1953), p. 101.

[5] Musson and Robinson, *Science and Technology*, p. 72.

[6] *Ibid.*, pp. 67–72.

'Mrs. Opimian: You have omitted accidents, which occupy a large space in the newspaper. If the world grew ever so honest there would still be accidents.
'Rev. Dr. Opimian: But honesty would materially diminish the number. High-pressure steam boilers would not scatter death and destruction around them, if the dishonesty of avarice did not tempt their employment, where the more costly low pressure would ensure absolute safety.'[1]

At all times the limits to advance in engineering have been set to some extent by the metal-working industries and not merely by the inventors; even less so by the obstructions of the patent law. Watt had been faced by the problems of finding cylinders that ran true throughout their length, and pistons to fit them, of devising valves that did not leak, or gearing that did not slip, of discovering packing that did not set up impossible friction, and similar difficulties, so that at any time it is impossible to allocate the total responsibility for delays in engineering progress to any single factor. Above and beyond all other difficulties Watt had to train his own engine-erectors. Despite the unkind comments that Watt occasionally made about the mechanical abilities of his fellow-countrymen, the greater number of them proved to be Scots. Was this, in another sense, a case of Necessity proving to be the mother of Invention, or must we find in the Scottish system of education the explanation of all this engineering talent? Murdock, Lawson, Pierson, Perkins, Cameron, Logan, Muir, Rennie, Ewart and Brunton – what an array of talent, including three Fellows of the Royal Society, do we find here! In answer to an enquiry from Erasmus Darwin about steam engines, Watt made his own summary of the situation:

'Preserve the dignity of a philosopher and historian; relate the facts and leave posterity to judge. If I merit it some of my countrymen, inspired by the *Amor Patriae* may say: "*Hoc a Scoto factum fuit*".'[2]

---

[1] *Novels of Thomas Love Peacock*, ed. D. Garnett (1948), pp. 808–09. See also Charles Dickens, 'A few conventionalities' in *Household Words*, 28 June 1851. For more information about boiler accidents and efforts to reduce them, see W. H. Chaloner, *Vulcan; The History of One Hundred Years of Engineering and Insurance, 1859–1959* (Manchester, 1959).

[2] Watt to Darwin, 24 November 1789, Muirhead *op. cit.*, Vol. II, p. 236.

1. Dr. James Gibson to James Watt junior, *8 October 1834, Doldowlod.*

The following letter, from Dr. James Gibson to James Watt junior, was written in 1834 in response to an enquiry from the Frenchman, F. Arago, who was then preparing his *Historical Éloge of James Watt* (English trans., 1839). It reports a few anecdotes of Watt's boyhood, related by his first cousin, Mrs. Jane Campbell, who was nearly of Watt's age and grew up with him. It includes the original account of the famous tea-kettle incident. There has been a modern 'debunking' tendency to dismiss this story as mythical, but it was evidently true. (The kettle crops up again, moreover, later in Watt's life, as a boiler in a series of laboratory experiments: see below, pp. 39–40, and E. Robinson, 'James Watt and the Tea Kettle. A Myth Justified', *History Today*, April 1956). Too much previsionary importance should not, of course, be attached to it, but this letter certainly illustrates Watt's early scientific curiosity and mathematical ability.

'My dear Sir

'Your friend Mr. Ewart having mentioned to me when he called here the other day, that M. Arago was desirous of having any familiar anecdotes respecting your father which might tend to illustrate his early genius, I have thought of three which I heard from his and your relative Miss Jane Campbell about 4 years ago – When a boy at school he had not robust health, and was thought rather idle whereupon the School Master recommended him to be withdrawn for a time and sent to the country to recover his health – when he had been there for a short time his parents went out to drink tea with their friends and see their son – while the elders were at table the subject of this éloge was found sitting by the parlour fire holding a plate (which he had filched off the table), to the spout of the boiling tea kettle noticing carefully the distances at which the steam condensed when it came in contact with the cold plate – who can say but this gave him the first idea of the power and elasticity of steam? – This is not exactly the way Miss Jane Campbell told the story but the facts are the same – On another occasion his parents had gone out to enquire about him, and having left some directions about his studies were desirous to know whether he had been paying more attention, when the Lady of the house told them she did not know whether he learned his lessons any better or not, but one thing she knew, and that was that he had surely been reading a great deal, for in these long autumn evenings it was very common for him to begin to tell his stories, one after another, and they were all so interested in them that the family were sometimes deluded out of half their nights rest, and often convulsed with laughter.

'This absence from school gave rise to another anecdote which I certainly think illustrative of his genius in after life – he had been considered by his Master idle and his father feared he might be dull, and was therefore more urgent in his enquiries every time, as to his sons studies, which however he could not get any satisfactory account of from those he lived

22

with but on this last occasion he was told that his son really was a very idle boy for he sat all day making all sorts of strange figures on the floor with a piece of chalk which nobody in the house could understand – his father desired to see his work and to his infinite pleasure found he had correctly worked out a mathematical problem – going a little deeper into the matter he discovered the walls and floor of his bed room all over with arithmetic and mathematical figures. The inventor of the steam engine had unfortunately left his slate behind him in the school – it is scarcely necessary to say that his Father was satisfied.

'If you do not happen to know these things of your father and if they come under the description of anecdotes that M. Arago wants I am happy to have it in my power to give them – I have adhered to Miss Campbells way of telling them as nearly as I could

I am My dear Sir

Your Very Truly

JAMES GIBSON

12 o'clock Wednesday night

8th October

I have written this off in a great hurry – but you will perhaps be able to read it.'

2. 'Professor Robison's Narrative of Mr. Watt's Invention of the improved Engine versus Hornblower and Maberley 1796.' *MS. Doldowlod.*

Professor John Robison was an early associate and friend of Watt at Glasgow University, where he joined with him in intellectual discussion and scientific investigations and where he witnessed many of Watt's early experiments on steam. He was one of the first men to whom Watt confided his invention of the separate condenser and he was privy to Watt's early efforts to develop the improved steam engine in collaboration with Dr. John Roebuck. In later years, after he became a professor at Edinburgh, Robison remained a lifelong friend of Watt and in 1796, during the Hornblower and Maberley piracy case, he came to give evidence in London and wrote the following 'Narrative', now among the Doldowlod papers.

QUEST[ION] Sir my Clients inform me that you had an opportunity of seeing the first Steps of the Invention which is to be the Subject of Consideration for the Court – pray where and how long have Mr. Watt and you been acquainted, and what opportunities have you had of knowing what occupyd his Attention.

ANSWER My acquaintance with Mr. Watt began in 1758. I was then a Student in the University of Glasgow, and then Studying the Same which I now profess to teach – Natural philosophy. The University were then building an Astronomical Observatory. Mr. Watt came to settle in Glasgow as a Mathematical and philosophical Instrument Maker, and was employed to repair and fit up a very noble Collection of Instruments bequeathed to the University by Mr. McFarlane of Jamaica, a Gentleman well known to the Scientific World. Mr. Watt had Apartments and a Workshop within the College. I had, from my earlyest youth, a great Relish for the Natural Sciences, and particularly for Mathematical and Mechanical philosophy. I was eager to be acquainted with the practise of Astronomical observation, and my Wishes were much encouraged by the celebrated Dr. Simpson professor of Geometry, Dr. Dick professor of that philosophy, and Dr. Moor professor of Greek, Gentlemen eminent for their Mathematical Abilities. These Gentlemen brought me with them into Mr. Watts Shop, and when he saw me thus patronised or introduced, his natural Complaisance made him readily indulge my Curiosity – After first feasting my Eyes with the view of fine instruments, and prying into every thing I conversed with Mr. Watt. I saw a Workman and expected no more – but was surprized to find a philosopher, as young as myself; and always ready to instruct me. I had the vanity to think myself a pretty good proficient in my favourite Study and was rather mortifyd at finding Mr. Watt so much my superior. But his own high relish for these things made him pleased with the Chat of any person who had the same tastes with himself, or his innate Complaisance made him indulge my Curiosity, and even encourage my endeavours to form a more intimate acquaintance with him. I loung'd much about him, and I doubt not, was frequently teazing him. Thus our acquaintance began –

It was interrupted in 1758. I left the College for the Navy, where I was a Midshipman four years, and was present in some of the most remarkable Actions of that War. My health suffered so much by a seafaring life that I was obliged to give it up, much against my inclination, and return to my academical habits. I was happy to find Mr. Watt settled in Glasgow, as fond of Science as ever. Our acquaintance was renewed, I believe with mutual Satisfaction. For I had now acquired some knowledge. I had lived in the closest intimacy with the late Admiral Sir Charles Knowles – and had been a good time employed in Marine Surveys – I had been imployd by the Admiralty to make the observations for the tryal of Mr. Harrison's famous Time piece – in that my habits had been such, that I reckoned myself more on a par with Mr. Watt, and hoped for a closer acquaintance. Nor was I disappointed. I found him as good and kind as ever, as keen after the acquisition of knowledge and well disposed to listen to the information I could give him concerning things which had not fallen in his own way. But I found him continually striking into untrodden paths, where I was always obliged to be a follower. Our acquaintance at this time became very intimate, and I believe neither of us engaged far in any train of thought without the other sharing in it. I had had the advantage of a more regular Education. This frequently enabled me to direct or confirm Mr. Watts Speculations, and put into a systematic form the random suggestions of his inquisitive and inventive mind. This kind of friendly commerce knit us more together, and each of us knew the whole extent of the others reading and knowledge.

I was not singular in this attachment. All the young Lads of our little place that were any way remarkable for scientific predilection were Acquaintances of Mr. Watt, and his parlour was a rendezvous for all of this description – Whenever any puzzle came in the way of any of us, we went to Mr. Watt. He needed only to be prompted – every thing became to him the beginning of a new and serious study, and we knew that he would not quit it till he had either discovered its insignificancy or had made something of it. No matter in what line – Languages – Antiquity – Natural History – nay Poetry, Criticism, and Works of Taste – As to any thing in the line of Engineering, whether Civil or Military he was at home, and a ready Instructor – Hardly any projects such as Canals, deepening the River, Surveys, or the like were undertaken in the Neighbourhood without consulting Mr. Watt. And he was *importuned* to take the Charge of some considerable Works of this kind tho' they were such as he had not the smallest experience in. When to this superiority of knowledge, which every man confessed in his own line, is added the naive simplicity and Candor of Mr. Watts Character, it is no wonder that the Attachment of his Acquaintances was strong. I have seen something of the World, and am obliged to say that I never saw such another instance of general and cordial attachment to a Person whom all acknowledged to be their Superior – But this superiority was concealed under the most amiable Candor, and liberal allowance of merit to every man. Mr. Watt was the first to ascribe to the ingenuity of a friend things which were very often nothing but his own surmises followed out and embodied by another. I am well intitled to say this, and have often experienced it in my own Case.

But the Circumstance which made Mr. Watts Acquaintance so valuable to me was the trait of Character I have already mentioned. Every thing became to him a Subject of new and serious Study – Every thing became Science in his hands, and I took every opportunity of offering my feeble Aid by prosecuting systematically, and with the help of Mathematical discussion, thoughts which he was contented with having suggested or directed. I thus shared the fruits of his Invention, and with gratitude I here acknowledge my obligations to him for that strong relish which I then acquired for rational mechanics, and which I have cultivated with great assiduity and pleasure all my life. I also shared with Mr. Watt a good deal of that subsidiary knowledge which he acquired as so many stepping stones in his way to some favourite Objects. He learned the German language in order to peruse Leopolds Theatrum Machinarum – So did I, to know what he was about – Similar reasons made us both learn Italian – and so of other things – And I cannot here pass over another Circumstance which endeared Mr. Watt to us all – He was without the smallest wish to appropriate knowledge to himself, and one of his greatest delights was to set others in the same road to knowledge with himself. No Man could be more distant from the jealous concealment of a Tradesman, and I am convinced that nothing but the magnitude of the prospect which his improvement of the Steam Engine held out to him and his Family could have made Mr. Watt refuse himself the pleasure of communicating immediately all his discoveries to his Acquaintances. Nay he could not conceal it. For besides the frankly imparting it to Dr. Black, and two or three more intimate friends, he disclosed so much of what he had been doing, that had it been in London or Birmingham, I am confident that two or three patents would have been expeded for bits of his Method by bustling Tradesmen before he thought himself intitled to solicit such a thing. And I have that confidence in the Native honour and modesty of his character that I do not believe that the contagion of plagiarism and Rivality which he has lived in for thirty years past has ever warped his mind so far as to make him assume to himself any thing that he thought the *serious* invention of another. I know indeed instances of his having made up forms of things which had been introduced by others *without thought*, and of which they did not perceive the proper situation or the advantage which knowledge like Mr. Watts could derive from them – as I see many of Mr. Watts Contrivances copied by ignorant Creatures who call themselves Engineers, merely because they are Watt's, and therefore must be good, tho' they are as often hurtful, by being improperly employed.

I doubt not but all this will be looked upon by some as mere panegyric. The ignorant are insensible to the pleasures of Science and have no notion of the Attachments which this may produce, and the low bred minds whose whole thoughts are full of Concealment, Rivalship, and Moneymaking, can hardly conceive a Mind that is not actuated by similar propensities. But I have a better opinion of those on whose feelings and Judgment the Issue of this Cause is to depend. I wish to show these Gentlemen what were my opportunities of seeing the Steps by which Mr. Watt arrived at his final discovery, and I am not afraid that they will misinterpret the Satisfaction I feel in having this opportunity of expressing my Sentiments of attachment to Mr. Watt. There is perhaps but one other person now alive, who was a witness to every Step of the Invention, and I regret exceedingly that his extreme illness makes it impossible for my Friend to avail himself of his Testimony. The thoughts of doing him an essential service have supported me in my journey hither under considerable Suffering, and when I find that not only the fortune but the fair Name of a Most worthy Man is concerned, I think that nothing less than life could have excused me from the sacred duty of every good Citizen, the support of eminent talents and worth against the vile aspersions of low bred and ignorant pretenders.

QUEST[ION]   The Court Sir wish to hear you give a short narrative of the Invention.
ANSWER   I think it was in the Summer of 1764, or perhaps in the Spring of that year that the professor of Natural philosophy in the University desired Mr. Watt to repair a pretty Model of Newcomen's Steam Engine. This Model was, at first, a fine play thing to Mr. Watt, and to myself, now a Constant Visitor at the Workshop. But, like every thing which came into his hands, it soon became an Object of most serious Study. This Model being an exact copy of a real Engine, the Motion of the Piston behoved to be the same, and the Strokes to be much more frequent. In consequence of this the Boiler was unable to supply more than a few strokes. The boiler was made to boil more violently. But this instead of continuing the motion by a more plentiful supply of Steam, stopped the Machine altogether; and we attributed this to the statical resistance to the entry of the Injection, which came from a height not much exceeding a foot. The Injection Cistern was placed higher, but without Effect – It was long before the true Cause was thought of, and in the mean time many observations were made on the performance. Mr. Watt had learned from Dr. Black somewhat of his late discovery of the latent heat of Fluids and of Steams. The doctor had established his doctrine by means of incontrovertible Experiments in the Case of Congelation and liquefaction, but had not yet devised any very simple and popular Experiments for showing the much greater quantity of heat which is contained in steam in a latent State.

But the great variety of curious and abstruse phenomena which were explicable by this branch of the theory made it a subject of much conversation among the young Gentlemen at College. Mr. Watt was one of the most zealous partisans of this Theory, and this little Jobb of the Model came opportunely in his Way, and immediately took his whole attention. [*Note in Watt's hand inserted on opposing blank page:* Dr. Robison is mistaken in this. I had not attended to Dr. Blacks experiment or theory on Latent heat until I was led to it in the course of experiments upon the Engines when the fact proved a stumbling block which the Dr. assisted me to get over. JW] He made many curious experiments, and Dr. Black publicly acknowledges in his lectures his Obligations to Mr. Watt for the chief experiments by which he illustrates and supports his Theory. I had not yet studied Chemistry, and Mr. Watt was my first Instructor. My Mind was fired with the inexhaustible fund of Instruction and Entertainment which I now saw before me, and I was more assiduous in my attendance on Mr. Watts occupations than ever, and studied the little Model as much as he did. He very early saw that an immense quantity of Steam was wasted. The great heat acquired in an instant by the Cylinder, by the admission of a few grains of Water in the form of Steam, was an incontestable proof of the great quantity of fire contained in it. And as this could come only from the Coals, Mr. Watt saw at once that the chief improvements that the Engine was susceptible of must consist in contrivances for increasing the production of Steam and for diminishing its waste. He greatly improved the Boiler by increasing the Surface to which the fire was applied – he made flues through the middle of the Water – he placed the fire in the middle of the Water, and made his Boiler of Wood, as a worse conductor of heat than the Brick work which surrounds common furnaces – He cased the Cylinder and all the conducting pipes in Materials which conducted heat very slowly – he even made them of Wood – After much acquaintance with his Models (for he had now made others) he found that there was still a prodigious and unavoidable waste of Steam and fuel, arising from the necessity of cooling the Cylinder very low at every effective Stroke – and he was able to show that more than three fourths of the whole steam was thus condensed and wasted during the ascent of the piston (subsequent experiments made with better apparatus showed him that the waste was much greater than this). I had seen all these contrivances and many of the Experiments, and had sometimes contributed my mite to lessen the expensive waste. But this great Cause of loss seemed to be unavoidable. At the breaking up of the College (I think in 1765) I went to the Country. About a fortnight after this I came to town, and went to have a Chat with Mr. Watt and to communicate to him some observations I had made on Desaguiliers and Belidor's Account of the Steam Engine. I came into Mr. Watts parlour without Ceremony, and found him sitting before the fire, having lying on his knee a little Tin Cistern, which he was looking at, I entered into conversation on what we had been speaking of at last meeting, something about Steam. All the while Mr. Watt kept looking at the fire, and laid down the Cistern at the foot of his Chair – at last he looked at me and said briskly 'You need not fash (trouble) yourself any more about that Man, I have now made an Engine that shall not waste a particle of Steam. It shall all be boiling hot, aye and hot water injected if I please'. So saying Mr. Watt looked with Complacency at the little thing at his foot, and seeing that I observed him he shoved it away under a table with his foot. I put a question to him about the nature of his Contrivance. He answered me rather dryly. I did not press him to a further explanation at that time, knowing that I had offended him a few days before by blabbing a pretty contrivance which he had hit on for turning the Cocks of the Engine. I had mentioned this in presence of an Engine-Builder who was going to erect one for a friend of mine, and this having come to Mr. Watts Ears, he found fault with it.

I was very anxious however to learn what Mr. Watt had contrived but was obliged to go to the Country in the Evening. A Gentleman who was going to the same house said that he would give me a place in his Carriage, and desired me to wait for him on the Walk by the River side. I went thither, and found Mr. Alexander Brown, a very intimate Acquaintance of Mr. Watts, walking with another Gentleman (Mr. Craig Architect). Mr. Brown immediately accosted me with Well have you seen Jamy Watt? Yes. He'll be in high spirits now with his Engine, is'n't he. Yes said I very fine spirits. Gad says Mr. Brown the Condenser's the thing keep it but cold enough and you may have a perfect vacuum whatever the heat of the Cylinder. The instant he said this the whole flashed on my mind at once. I did all I could to encourage the Conversation, but was much embarassed. I durst not appear ignorant of the apparatus, lest Mr. Brown should find that he had communicated more than he ought to have done. I could only learn that there was a Vessel called a Condenser which communicated with the Cylinder, and that this Condenser was immersed in cold Water and had a Pump to clear it of the Water which was formed in it. I also learned that the great difficulty was to make the piston tight, and that leather and felt had been tried and found quite unable to stand the heat. I saw that the whole would be perfectly dry, and that Mr. Watt had used Steam instead of Air to press up his Piston which I thought by Mr. Browns Conversation was inverted – We parted and I went home a very Silent Companion to the Gentleman who had given me a seat – Next day, impatient to see the Effects of the separate Condensation, I sent to Paisley and got some tin things made there, in completion of the notion that I had formed. I tryed it as an Air Pump, by making my Steam Vessel communicate with a Tea Kettle, a Condenser, and a Glass Receiver. In less than two minutes I rarifyd the Air in a pretty large Receiver more than 20 times. I could go no further in this process because my Pump for taking out the Air from my Condenser was too large and not tight enough. But I saw that when applyd to the mere purpose of taking out the Air generated from the Water, the Vacuum might be made almost complete. I saw too (in consequence of a Conversation the preceding day with Mr. Watt about the Eduction Pipe in Beighton's Engine) that a long sink pipe or syphon would take off all the Water – In short I had no doubt that Mr. Watt had really made a perfect Steam Engine.

I was very unfortunate in two visits I made to Glasgow during that Summer, Mr. Watt being from home, once at Greenock seeing his Father who was ill, and the other time on a Survey for a Canal. When I came to Town for the winter, I found that Mr. Watt was again from home, and that he was deeply engaged with his Engine. His situation in life made it imprudent to engage in great expences, and he was obliged to look out for an Associate. Most fortunately there was in the neighbourhood such a Person as he wished. Dr. Roebuck, a Gentleman of very uncommon knowledge in all the branches of Civil Engineering, familiarly acquainted with the Steam Engine, of which he employed several on his Collieries, and deeply interested in their Improvement. He was also well accustomed to great Enterprises, of an undaunted Spirit, not scared by difficulties, nor a niggard of Expence. Such a Man was indispensably necessary to one of Mr. Watts Character, modest, timid, easily frightened by Rubs and Misgivings, and too apt to despond. I do not know who pointed him out to Mr. Watt. He was well acquainted with Mr. Watts Talents and admired them. I believe the Connection was very soon formed. Dr. Black and all Mr. Watts friends were happy at seeing so fair a Commencement. At this time I had not the pleasure of being known to Dr. Roebuck.

I think it was the middle of Winter before I saw Mr. Watt. When we met, he most frankly told me all his contrivance, and I took Care to receive it all as perfectly new to me, that I might not commit Mr. Brown. I remember well that when he complained of the great power expended in working pumps sufficiently large to exhaust a Condenser even of moderate size (because they must do it at one Stroke, against the whole pressure of the Atmosphere) I mentioned the Observation that I had formerly made to him on the Eduction pipe of Beighton's Engine, and the contrivance which I would deduce from it for clearing the Condenser of Water, Mr. Watt said 'O Man do you imagine me so dull as not to have thought on that long ago – But I could give you many reasons why it will not answer as well as a pump. I wish I could as quickly get quit of the Air as of the Water without a Pump. I dont despair even of this.' He now informed me of many curious properties of Steam relative to its heat and elasticity, explained his methods of condensation, mentioned some remarkable facts relative to this Subject, which pass to this day before the Eyes of every body, without being noticed or understood by hundreds who call themselves Engineers and builders of Steam Engines.

After this time my meetings with Mr. Watt were less frequent. He was much from home, working with his Engine. And I was now obliged to devote my whole attention to another Subject. Dr. Black, my professor in Chemistry, was now removed to Edinburgh, and by his very unmerited recommendation the University of Glasgow placed me in the Chemical Chair which He had just quitted. Frightn'd by my own good fortune I was obliged to strain every nerve to do some Credit to so partial a Recommendation, and I was obliged to relinquish all other occupations of my thoughts. But I had now learned all the principles of Mr. Watts Invention, tho' I had never seen either his Engine or any Model or drawing of it. I knew of his employing Steam in place of the Atmosphere to press forward his Piston, altho' it was long ere I knew the way in which he introduced it. I thought he simply admitted it from the surrounding Case by the open Mouth of the Cylinder, and it was not till I was in St. Petersburgh that I learned that he also introduced it (without a Steam Case) by a pipe. This however was a *natural* part of the leading thought, and indeed was practised by him in his very first experiments. In this experiment (which was made with a common Anatomists great Injection Syringe for a Cylinder) the Piston Rod passed through a Collar of Leather in the Cover of the Cylinder, and the Steam was admitted thro' another aperture on the same Cover and it escaped into the Condenser by a similar aperture in a Cover on the other End of the Cylinder. Long after this I found that the little Apparatus which I saw on his knee, and which he shoved under the table with his foot, was the Condenser in this first Experiment. I discovered that I had not comprehended the whole Contrivance so completely as I imagined – But tho' I was ashamed of my ignorance, my vanity would not let me acknowledge it, and I took circuitous ways of learning more exactly the precise State of the Engine. I was living in Edinburgh during the Summer 1767 near Dr. Black in order to prepare myself for my arduous Task, and in my Conversations with Dr. Black I frequently introduced Watts Steam Engine. I one day asked him why Mr. Watt never thought of impelling the piston by Steam much stronger than common Air, mentioning the way in which I would introduce and manage it. He then corrected me in some parts of my proposed Construction, and described Mr. Watts with accuracy, and bade me reflect on the enormous size and Strength which must be given to the Boiler, and the expence of fewel in supplying steam so dense and so hot. All this I had thought on already, and only wanted to learn what he had just now told me. And now I am fully entitled to say that in the Summer 1767 the whole contrivance was perfect in Mr. Watts Mind, altho he had neither executed the double Stroke nor that most beautiful contrivance of cutting off the Steam before the Piston reached the bottom of the Cylinder, a contrivance which in a moment fits the Engine, however great and powerful, to any the most trifling Task, and makes it more manageable than any other Engine whatever that is not immediately activated by the hand of Man. Indeed any person who deserves the name of Engineer must see, and if he speak from the Conviction of his Conscience must acknowledge, that the whole Contrivance was perfect in Mr. Watts Mind in his very first Tryal.

During the two following Winters, notwithstanding Mr. Watts frequent Absence from Glasgow, and my constant occupation with my chemical Lectures, I had many opportunities of conversing with him, and learned all his difficulties and embarassments. He struggled long to condense with sufficient rapidity without Injection, and exhibited many beautiful Specimens of ingenuity and of fertility of resource. Many pretty Schemes occurred to him for a rotatory Engine. Some of these I am sorry to find that he has neglected. I am confident of their complete Success, and tho' I agree with him in thinking that his Engine with a double Stroke is superior to them all, I should have been glad that they had been executed, because they would have given a most brilliant Specimen of his wonderful Ingenuity and of his knowledge – For indeed the management of Steam to perfection is the employment of an accomplished philosopher, and far beyond the Ken of a Pump or Water Closet maker. Mr. Watts private Notes must contain a treasure of curious knowledge, and it is a thousand pities that this cannot be laid before the public without raising an Host of Pirates and plagiarists who would pester the patent office with wonderful Inventions from every quarter. I see many of Mr. Watts contrivances which I know to be founded on his perfect acquaintance with the subject, copied by ignorant Tradesmen, merely because they are Boulton and Watts and then introduced to their hurt. Such is the deference with which these Persons in their Conscience look up to my Friend's superior Skill. And yet (such is the power of Avarice) these Creatures will say that Boulton and Watt steal their wonderful Inventions! I cannot resist the present impulse to give a proof how unlikely these Gentlemen are to dirty their fingers with such petty Larceny. In the Summer of 1768 I took a Ride thro' some parts of England. Mr. Watt gave me a Letter of Introduction to Mr. Boulton, with whom he by that time was a little acquainted. Mr. Boulton received me with the politeness that is natural to him, and that regard which he knew to be due to one whom Mr. Watt called his intimate Friend. At dinner, in company of several Gentlemen well known to men of Science I spoke of Mr. Watts discoveries respecting the management of Steam, and of his Intention of availing himself of them for improving the Steam Engine, under the protection of a Patent. Mr. Boulton joined in praising Mr. Watts Ingenuity – After dinner Mr. Boulton kindly took me thro those parts of his magnificent works that I had not yet seen. As we walked thro' a Yard he drew me aside and said does Mr. Watt seriously mean to apply for a Patent. I answered Yes, most certainly, in conjunction with Dr. Roebuck – Why then said he I will stop that work, pointing to some brick pillars which lay on the other side of a large Water Course or Ditch. I am about erecting a Steam Engine, very unlike what you described, but where I should have availed myself of what I learned from Mr. Watts Conversation. But this would not be right without his Consent. Are these the Men who would pilfer the inventions of others.

QUEST[ION] As you profess a predilection for mechanical performances I may presume that your acquaintance with what has been done in this line is pretty extensive. Have you ever seen in the Course of your Observation or Reading, the distinguishing principles which Boulton and Watt suppose are secured by patent, applied to these or similar purposes before the Year 1769.

ANSWER Never – and I am almost certain that they never were – during my Residence in Glasgow, I was in the habits of continual Intimacy with Mr. Watt. All who know him know that it is his greatest pleasure to communicate his knowledge to those who have a relish for it. I have reason to think that he never from any kind of Jealousy, concealed any thing from me. From the day that he (I may almost say we) began to play with the College Model, I knew almost every Step of his thoughts. He was confined to his Business. I was more at large, and going about the College. I ransacked the Libraries for every Book that he wanted – and every quotation which he met with made him impatient till he got at the Original. I saw every book that he got by any other Channel besides the public Libraries. So I may safely say that I knew the whole extent of his Reading. Our abode was too far out of the Circle of Business for allowing us to be informed of the numberless projects that are every day born and buried in this busy Country. I can say with great Confidence that nothing ever ocurred to Mr. Watt, either by Reading or information, of his leading principle of a Steam Vessel perpetually and universally hot – All the other Contrivances of Separate Condenser, Air and Water Pumps – Amalgam or Rosins or Fats for keeping the piston air tight, are but so many emanations from this first thought and I will venture to say, they all came into his mind in succession, and nearly in the order I have stated, after he said to himself 'Let me make an Engine working by a Piston in which the Cylinder shall be continually hot and perfectly dry'. I will venture to say that in no book previous to that date is there any Account or proposal of such a thing, if we except some attempts to put the Steam Vessel of Worcester or Savary's Engine in this predicament, by means of a travelling Mass of Oil or Air which was to be interposed between the Steam and the Water that was to be raised – Of these Mr. Watt and I had some very imperfect Account – but they never interested him, because the very nature of the operation made it impossible to do any thing more than approximate to the desired Object – After the true Method had occurred to Mr. Watt viz by the help of a Separate Condenser, he did not think Savary's Engine below his Notice, and he showed me some ingenious constructions for this purpose, when the whole height was divided into Lifts, and something equivalent to a Separate Condenser was applied. Some parts of these Constructions (if not transpired from Mr. Watt) have since that time occurred to others, such as Mr. Blakey, Mr. Kerr etc., who have exhibited them to the public as great Inventions, the common language of ignorant persons, and used by them bonâ fide, because they know nothing better – But nothing of all this has any applicability to an Engine working by a Piston, and they were all posterior (even in the mind of Mr. Watt) to the Method for which this patent was granted.

I must say further that the thought was wholly Mr. Watt's. For this I have every authority that can be wished for – I am certain that when I went out of town (in May I think 1765) he had not thought of the method of keeping the Cylinder hot – and I am as certain that a fortnight after, he had completed it and confirmed it by Experiment. Dr. Black, the first philosophical Chemist of his time, and the most scrupulous Man upon Earth with respect to Claims of Originality, gave this to Mr. Watt in the most unqualified terms the first time I saw him, after I had learned it from Mr. Brown, and long before I saw Mr. Watt and got it more distinctly from himself. I believe that Dr. Black was the chief means of forming the connection between Mr. Watt and Dr. Roebuck, and I recollect most distinctly his saying to me that Watt would have some difficulty in managing Dr. Roebuck, who at that time had not become a complete Convert to the doctrine of latent Heat – Accordingly it was so, and Mr. Watt was obliged to yield for some time to the doctors Confidence in his own great Experience. The doctor thought to produce the Condensation with sufficient rapidity and accuracy by a very extensive surface, and Mr. Watt knew that it also required a great quantity of Water or other matter to receive the emerging heat. I know that these differences of opinion retarded the completion of the Engine. But doctor Roebuck had too much judgement not to see the conclusiveness of the Experiments by which the doctrine of latent heat is established and not to yield to their force, and every thing at last went to their mutual Satisfaction. Dr. Roebuck knew Mr. Watts Talents and most liberally praised them. His timidity, his disposition to despond when under unforeseen difficulties, and his painful anxiety and diffidence in himself were frequently the subjects of friendly merriment at the doctor's fireside, and I have often heard him say that without his help, and even his instruction in many points of the Construction, Mr. Watt could never have gone on. I have even heard him mention some important but subordinate parts of the Engine which were of his Contrivance. But I never heard him lay the smallest Claim to the leading thought of a hot and dry Cylinder for the Piston to work in, and therefore a Separate Condenser. I never heard him call it *my* engine, nor *our* Engine, but uniformly *Watts* Engine, when he had occasion to speak of it as distinct from the *Old* or *Newcomen's* Engine. I remember Mrs. Roebuck saying one Evening Jamy is a queer Lad, and without the doctor his Invention would have been left, but Dr. Roebuck won't let it perish. I mention all these trifling things because I have often heard Gentlemen living in the Neighbourhood of Borrowstonness speak of this new project as Dr. Roebucks, in which he was assisted by one Watt from Glasgow. One Gentleman in particular Mr. Graham of Airth, insisted with me that Dr. Roebuck was the Inventor. But one day Mr. Graham came home from Falkirk, where he had seen Dr. Roebuck and engaged him in Conversation on the Subject. He told me that he now saw plainly that Mr. Watt was the sole Author, and said that he would be at some pains to undeceive some Gentlemen of the Neighbourhood who were of the opposite opinion. This was very natural. Dr. Roebuck was a Gentleman of uncommon knowledge in every thing of this kind, and considered as the first judge in all that Country of all such Matters, whereas Mr. Watt was an entire stranger, and a stranger from Glasgow, by no means a recommendation in that Country. I remember also that in 1774 or 1775 [*Note in Watt's hand:* This date must be wrong. Mr. Watt's Act of Parliament was obtained in May 1775, and he did not enter into Partnership with Mr. Boulton until later in that Year. It was several Years after this, before Boulton and Watt's profits could form a subject of regret, or envy to Dr. Roebuck. Moreover the Dr. had been under more pecuniary obligations to Mr. Watt during their Connection, than Mr. Watt to him, and it was chiefly his pecuniary embarassments which caused so many (5 Years) of the Patent to elapse, without any progress being made in the introduction of the Invention for the outlays he had incurred, which were not considerable.] after my return from Russia I had some Conversation with Dr. Roebuck. The doctor spoke with some dissatisfaction of Boulton and Watt. They were now, he said amassing fortunes from a project which his Misfortunes had obliged him to cede to them. They seemed to have forgotten that he had suffered all the Anxieties attending the infant project, he had run all the risk – and the Risk had been very great, both from the novelty of the thing and from Mr. Watts delicate health and his timidity under difficulties. That without his continual encouragement and support it never would have succeeded. He had ceded his right on very moderate Terms, and he had expected some remembrance of this. In this disposition to repine at an opportunity which he had lost of benefiting himself it would have been most natural for Dr. Roebuck to put a high value on any part that he had had in the discovery – and I listened with some anxiety to hear if he advanced any Claim of this kind. For I knew that any such thing from Dr. Roebuck would be received with much deference. But I have the most distinct recollection that he made no Claim whatever of this Sort, but, on the Contrary, spoke in the highest terms of Mr. Watt's ingenuity and inexhaustible Resource of Invention.

The duties of my profession call my Attention to a great variety of very interesting Objects. Of all these my favourite Object is practical Mechanics. I have therefore hunted every where for information, And my opportunities have been considerable. Understanding most of the Languages of Europe, I have looked into almost every book which treats of such things; And, in particular, I have searched for every project in mechanics, description of machines, and Schemes of Public Works. I can recollect but one trace of any thing like a Separate Condenser of Steam. This is in a volume of the Commentarii de Rebus in Medicina et Scientia naturali gestis. I cannot now recollect the volume, and only remember that it is a late one (indeed this whole work is of a date posterior to 1769). In this volume there is a short account given of an Air Pump by Mr. Wilcke of Upsal in Stockholm, precisely such as I made when I heard of Mr. Watt's contrivance. It is mentioned as a thing which the Reviewers had forgotten in its proper time and they say '*dudum fabricavit*'. I mentioned this about a year ago to Dr. Black when we were speaking of some curious Observations of Mr. Wilcke on the Cloud which appears in the Receiver of an Air pump when damp Air is suddenly rarify'd. The doctor told me that when he was yet in Glasgow he had a pupil of the name of Williams or Williamson, from the Mine College in Sweden. That this person was intimately acquainted in Dr. Roebucks family, and, he believed, also with Mr. Watt. That he was in this Country almost three Years, and fully understood all his theory, and he had no doubt but that Dr. Wilcke owed to him all that he had published on that subject. He thought it equally probable that this project of an Air pump had transpired in some of our Conversations, it being a thing in which we put no value.

Nor have I seen any thing of a Piston impelled by Steam. In the strange and crude Accounts which Papin gives of his Steam Engine in the Acta Eruditorum there is mention made of a Wooden Piston. But this is merely used to fill up a space, and to prevent the Steam from coming into Contact with the Cold Water. It should more properly be called a *Float*, and its sole office is to fill up a space and to diminish this Contact of Steam and cold Water. Papin seems to have known that this would waste Steam – Many such floats have been proposed for Savary's Engine – In another Paper Papin describes a Piston impelled by fired gunpowder – and says that the same thing might be done 'per vaporis aquae ebullientis' – In another paper he mentions a piston which is impelled by Steam through a certain part of the Barrel, when it opens a hole, thro which it escapes – But in all this there is not any similarity either in purpose or Effect between this and Mr. Watt's Invention. The Steam is not substituted for common Air – Nor is it condensed; nor does the piston perform the office of a first Mover in any Machine whatever.

These are all the traces that I can find of any thing like what is done in the Engines constructed after the Method of Boulton and Watt, and I therefore think myself well intitled to say that the principles are solely of Mr. Watt's Invention.

QUEST[ION]   Do you think that an Engineer who can erect a Newcomen's Steam Engine could be able merely by reading the Specification given by Boulton and Watt, to contrive and erect an Engine constructed on their principles, and which like theirs would make a great saving of Steam, and consequently of fuel?

ANSWER   I find greater difficulty in answering this question. Many call themselves Engineers who do not deserve the name of Mill Wright or Pump Maker. I would call that Man only an Engineer who knows the mathematical principles of Mechanics Hydrostatics and Hydraulics, at least so far that he can calculate the power of his Engines as they are modified by their construction. I see some who make a great figure in a Work called the Repertory or Repository, who show considerable ingeniuty in many little contrivances and use many long legged words in their descriptions, and would be very much dissatisfyed with any other name than Engineer. Yet I am certain that they are totally ignorant of mechanical or mathematical principles, and incapable of demonstrating the simplest of all hydraulic propositions, and the foundation of all the rest, namely how much water will run thro' a square inch placed one foot for instance under the Surface. Nay I know an Engineer, a pupil of Boulton and Watt, who has erected several very good Newcomen's Engines in the Neighbourhood of Glasgow, so completely ignorant of principle that when he erected a forcing pump which cost near £300 at a Cotton Mill near Paisley, he inserted the Main or Rising pipe at the very top of the Air box. I observed the Jets wabbling, and keeping time with the Strokes of the Pistons, and imagined the Air Box to be very small, and was surprised to see it above two hogsheads. I then thought that it had let the Air leak out by some Crevice. Not finding any, I sent for the drawing of the Pump, and found it as above stated, and when I questioned Mr. Muir about it, he said he had always made them with the Main rising either from the top or from the side, as suited best, and that his works had always given satisfaction to his employers. As his Credit was concerned in this one, he would come out and correct it. He came in my absence, and changed his four inch Barrels for six inches, and increased their Stroke – but made no change about the Air box – and there was no change in the performance but the bursting of all the leather pipes by the unequal Supply of the Pumps. Yet this man showed some ingenuity in several things where principle was not concerned.

I cannot pretend to say what ingenious vagaries such Engineers might run into after reading Boulton and Watts Specification. Perhaps they would make a nice and warm Water Closet. Nor am I a competent Judge what an Engineer of ordinary Intelligence would make of the Specification. I know too much of the Machine and of its principles. But, divesting myself as much as in candour I think I should of this advantage, I think in my Conscience that an Engineer acquainted with the principles of Newcomen's Engine, as far as they were understood about the year 1769, who risks *his own* money upon it, and will therefore think seriously about it, *cannot fail* of erecting an Engine which will certainly work, and make a great saving of Steam and Fuel. I suppose him ignorant of the doctrine of latent heat, but having as much common sense as to know that all the heat of the steam comes from the Fuel, and that every foot of Steam costs Money. I further suppose him so far acquainted with the mechanical and hydrostatical principles of the Engine as to know that the Steam does not force up the piston, but only allows it to be dragged up. I suppose him to know that the Cylinder will condense Steam as long as it is colder than boiling Water, and that the piston will not rise till all is hot and dry, and that more Steam will be condensed as the piston rises. And I suppose that he believes that the Steam will rush out of the Cylinder into any cold place that is totally empty, even of Air. With these pieces of preparatory knowledge or belief, I think it impossible for the Engineer to avoid making an Engine that will do the Work of Newcomen's with much less Fuel. And I think that the Specification not only tells him what things or parts his Machine must contain, but also puts him into a train of thought which suggests to him their forms and their arrangement. His Cylinder must be as hot as boiling water. It must therefore be dry. Therefore he cannot use Water to make his piston tight, for it would, or *must* boil off. He is told that greasy matters will do, and taking this for a truth, he uses them. Fact shows that the piston will be tight, if the Cylinder is well bored. But how shall the Steam be condensed if the Cylinder is boiling hot? He is told 'in another Vessel'. Stupidity itself will see that there must be a communication between them, and that this communication must not be perpetual, but must be cut off when the piston is to be allowed to rise by the action of the Outweight. As to the mode of condensation, the one he is accustomed to, viz, by injection, will readily present itself – perhaps some Engineer more than usually ingenious, or disposed to refinement, may think of doing it by the mere cold surface of the Condenser. This person will find more difficulty. But his first tryal will show him that tho it does it perfectly, it will do it too slowly, and he will content himself with the vulgar method of Injection. Nor is there any thing abstruse in the manner in which this injection shall be produced. Let him copy the old injection, and he will not do amiss, tho' not the best possible – But how shall he get quit of the Water and Air that are generated or thrown in. He is told 'by pumps'. Hearing that he is to pump out Air, he will naturally try to make a good pump – And *any* good pump will do. Nor is there any difficulty how he shall place this pump – Surely in the lowest part of the Condenser, where the Water naturally lodges – His acquaintance with the old Engine will give him some Notion of the quantity of Water generated and injected at each stroke. All this must be taken out by one Stroke of his pump – Common sense will tell him that his pump must not be too small for this, and that a larger pump will do no harm, but rather good, by taking out more Air – And, having it in his choice to have his pump piston either at the top or bottom of the barrel when the Steam is to be let into the Condenser, common sense will make him choose to have it at the top, that the steam may be drawn in (as is usually said) by Suction. Following these obvious and natural Suggestions of the Specification and of common sense, an ordinary Engineer cannot avoid making an Engine on the principle of Boulton and Watt, and this Engine cannot fail of saving much Steam and Fuel – a very moderate degree of further knowledge of the physical properties of Steam will teach the Engineer that a small injection will be more heated that this will leave a stronger Steam, and this will admit of a smaller load on the Engine. This Engineer will make a much more perfect Engine, and careful attention to his *own Engine*, not ignorantly copied, but half invented, cannot fail of making him capable of improving his Engine, and of making one much superior to those of another Artist, who has copied one of Boulton and Watts.

But I can give the Court a proof from fact that much less information than is given in the Specification is sufficient for enabling an intelligent Engineer to erect an effective Engine, or to comprehend Mr. Watts Principles. When I was in Russia, Inspector of the Imperial Marine Academy, I recommended to the Admiralty College the employing a Steam Engine to drain their great dock Bason, instead of two very expensive Windmills, which had enabled them to dock only three Ships from the time of their erection in 1726. I wrote to Mr. Watt on the Subject, that I had communicated plans of a Newcomens Engine, by which I had engaged to drain their Bason in 11 days. But as I was convinced of the great superiority of the Engine which I left him occupied with in 1770, I wished him to undertake it because this would still more justify my project in point of Œconomy – He returned me for answer that I was fully able to do much more than I had engaged, and that as the execution of such a project would do me credit, he declined interfering (I desire the Gentlemen of the Jury to judge whether this was the behaviour of a grasping Tradesman). The day that I received this Letter I went to drink tea with Mr. Model the Court Apothecary, one of the first Chemists of the Age, with whom I had frequently spoken of this project, and whom I had instructed in the doctrine of latent Heat. I found sitting with him Mr. Aepinus a Gentleman no less eminent for his beautiful theory of Magnetism and Electricity. I mentioned Mr. Watts genteel declinature, and also a passage of his Letter in which he said that by admitting Steam to press down the Piston its want of perfect tightness was not so hurtful as appeared at first Sight, because the Steam which got past would only be lost, but would not choke the Engine. Model broke out into an exclamation confirming what I said. Aepinus did not see the force of what we said, and Model took out his pencil to make a Sketch which would explain it to him. Not readily finding a bit of Paper, I pulled a Bit out of my pocket on which he made a Sketch. This happened to be an Official Report which I had that day received at Cronstadt, and which I kept with many things of this kind and they came home with my other papers. I submit it to the inspection of the Court, and presume it will be acknowledged as a convincing proof that Mr. Model completely understood Boulton and Watts Method.

I have another Fact on which I found my opinion of the Sufficiency of the Specification. George Houstoun Esq. of Johnston is a Country Gentleman of Rank and Fortune. He has had a Newcomen's Engine these 29 Years, and is perfectly acquainted with its performance. He understands its principles as far as any Man who is totally void of Mathematical and Philosophical knowledge. Having that distrust of my own competency to answer this question which I have already mentioned, and thinking Mr. Houstoun a proper Subject for this sort of Experiment, I cut the Specification out of a Copy of the Act of Parliament and sent it to him, telling him the Nature of the question in debate, and requested that he would consider it seriously, and see whether he understood Mr. Watts principle. If he did, I further desired that he would make a Sketch of an Engine constructed on this principle, and give me his reason for every part of the Construction. He had never seen an Engine of this kind, and tho' I had often mentioned it in his hearing, he never before heard its principle explained. The only piece of information which I gave him was that every ounce of Steam contained as much heat as would raise 1200 ounces of Water one degree all of which came from the Coals, and all of which will emerge when it is condensed into water – for this I desired him to trust my information. He returned the following Answer.

I submit this also to the Court as a proof that Mr. Houstoun was able to direct the Erection of an effective Engine on the principle of Boulton and Watt.

How rarely do we meet with such a Conjunction of Science and Art – how precious when it is found – how much then does it deserve to be cherished! – What advantages have been derived within these twenty years from this fortunate Union – How much then does it become our Courts to encourage and support it against the unprincipled attacks of ignorant and greedy plagiarists, who would deceive our Men of property, ruin them by expensive projects which terminate in disappointment – and thus discourage those who alone can by their Capitals give any effectual aid to the energy and genius of this Country! – We boast of our Newton, and sit him at the head of our philosophers – Our B and W want only justice, and all Europe will place them at the head of our Artists.

I know that it has been repeatedly objected to this opinion of Men of Science concerning the sufficiency of the Specification that Mr. Watts own Accounts are in opposition to it. He had to encounter many difficulties before he perfected his Machine, even after obtaining his Patent. I know this well. But this was chiefly in subordinate parts of the Undertaking. I firmly believe that the great principles were as perfect in his mind in a few hours as they are at this day, and that the physical parts of the problem were so completely solved by his first Model as they are now by his best Engines. But when Mr. Watt was engaged in bringing the Contrivance to perfection he wished to perfect every part. He also wished to make his Engine not only the best, but the cheapest in the World – He struggled long, in opposition to his own judgement at Dr. Roebucks instance to perform the Condensation without injection. He had a predilection for the Wheel Engine, and much time and labour were spent on it, while he was uncertain whether he should bring this or the reciprocating Engine first to the Market. He had experience to acquire in great Works, and in the practice of several trades employd in such Construction. He had Workmen to instruct, and to form, and to keep with him after they had acquired from him a little knowledge and were worth bribing away from him. But the chief cause of the delay was that indelible trait in Mr. Watts Character, that every new thing that came into his hands became a subject of serious and systematical study, and terminated in some branch of Science – Allow me to give an instance. A Mason Lodge in Glasgow wanted an Organ. The Office bearers were Acquaintances of Mr. Watt – We imagined that Mr. Watt could do any thing, and tho' we all knew that he did not know one musical note from another, he was asked if he could build this Organ. He had repaired one and it had amused him. He said Yes – but he began by building a very small one for his intimate friend Dr. Black, which is now in my possession. In doing this a thousand things occurred to him which no Organ builder ever dreamed of – nice Indicators of the strength of the blast Regulators of it etc. etc. He began to [build] the great one. He then began to study the philosophical theory of Music – Fortunately for me no book was at hand but the most refined of all, and the only one that can be said to contain any theory at all, Smith's Harmonics. Before Mr. Watt had half finished this Organ, he and I were completely masters of that most refined and beautiful Theory of the Beats of imperfect Consonances – He found that by these Beats it would be possible for him, totally ignorant of Music, to tune this Organ according to any System of temperament – and he did so, to the delight and astonishment of our best performers – in prosecution of this he invented a real Monochord of continued Tone – and in playing with this he made an Observation which, had it then been known, would have terminated a dispute between the first Mathematicians in Europe, Euler and d'Alembert, and which completely establishes the theory of Daniel Bernoulli who differed from both of those Gentlemen about the mechanism of the vibration of Musical Chords, and as completely explains the harmonic Notes which accompany all full Musical Notes, overturning the theories of Rameau and Tartini.

No wonder that the attachment to Mr. Watt was strong, when persons of every taste and every pursuit found in him an inexhaustible fund of instruction and Entertainment – Yet this is the Man who pilfers the paltry secrets of a Hornblower and a Bull. Credat Judaeus apella – non ego.

1 Portrait medallion of James Watt by George Mills. *By courtesy of Birmingham Art Gallery*

to the arithmetical mean betwixt the
two heats & that it was scarcely sensible
heated to the finger.
I took a ^beat Glass tube & inverted it on
the nose of a tea kettle
the other end being immersed in
cold water I found ~~a small increase of~~
~~water in the~~ on making the
kettle boil that tho there was only
a small increase of the water in
the frigeratory that it was became

3. Extracts from James Watt's 'Notebook of experiments on heat'. *MS. Doldowlod.*

The following document, an extract from Watt's own experimental notes, provides more evidence of his careful scientific procedure, of his repeated experiments, of his independent discovery of the phenomenon of latent heat, and of the assistance given by Dr. Joseph Black.

About 6 or 8 years ago My Ingenious friend Mr. John Robison having conceived that a fire engine might be made without a Lever by inverting the Cylinder and placing it above the mouth of the pit proposed to me to make a model of it which I set about but never compleated he going abroad and I having at that time little knoledge of the scheme. however I always thought such a Machine might be applyed to other as valuable purposes as raising Water. on making some experiments with a Digester on the force of steam I thought that in that way a machine for some such purposes might be made advantageously, and being employed to put in order a small model of a fire engine belonging to the natural philosophy Class and made by Jonathan Sisson I mett with considerable difficultys in the execution owing to the very bad Construction of some of its parts but having at Last overcome all difficultys And made it to work satisfactorily I was surprized at the immense quantity of fuel it consumed in proportion to the Cylinder which was only 2 Inches diameter and which I imputed to the heat Lost thro the metallic Cylinder.

[He then tried cylinders made of wood and began to experiment on improving the packing of the piston and to produce a different type of boiler.]

I had often observed that the best way of heating bodies was to bring them in Contact with the burning fuel. the great distance from the fire to the boyler in fire engines seemed in consequence to be wrong. after many fruitless thoughts on the subject I saw no boyler so perfect in that Respect as the common tea kitchen (an Invention for which we are beholden to the Chinese) here the fuel is always in Contact with the sides of the boyler containing the water. the outside may be of wood with this advantage that very little heat will be able to penetrate it. the Inside of very thin Iron which will considerably diminish the Expence and being constantly covered with water it cannot burn. this boyler I put in practice and its effects answered my expectation evaporating prodigious Quantitys of Water and consuming little fuel. It burned 4 lb. of coals per hour and in that time evaporated 20 lb. of water. . . .

. . . About this time I conceived the plan of making a Machine something like Capt. Savarys but to raise water to any height by suction only by making a series of Receivers one above another so that each would raise the Water to the sucking pipes of the other the steam to be conveyed to them thro copper pipes surrounded with chaff to prevent the escape of heat.

But before any thing could be known certainly about the best manner of constructing Fire engines there were several facts necessary to be determined.

D

[Watt then proceeds to describe a series of experiments on the expansion of steam, the consumption of fuel, the conduction of heat thro metals to water, etc.]

I found thus that the Quantity of water used for Injection in fire engines was much greater than I thought was necessary to cool the Quantity of water contained in the steam down to below the boiling point. I mixed 1 part of boiling water with 30 parts of cold water and found it only heated to the mathematical mean betwixt the two heats and that it was scarcely sensibly heated to the fingers

I took a bent Glass tube and inserted it into the nose of a tea kettle . . . the other end being Imersed in cold water. I found on making the kettle boil that tho there was only a small increase of the water in the frigeratory that it was become boiling hot. this I was surprized at and on telling it to Dr. Black and asking him if it was possible that water under the form of steam could contain more heat than it did when water He told me that had long been a tenet of his and explained to me his thoughts on the Subject and in about a week after he tryed the experiment with the Still.

4. 'Report by Messrs. Hart of Glasgow of conversations with Mr. Watt in 1817. Communicated by John Smith. 19 March 1845.' *MS. Doldowlod.*

More light is thrown on Watt's invention of the separate condenser, and on his early efforts at developing the engine with Dr. Roebuck at Kinneil, by the following report of conversations between Watt and Messrs. R. and J. Hart, of Mitchell Street, Glasgow, in 1817. This document, among the Doldowlod papers, is accompanied by a letter, dated 19 March 1845, from John Smith, of 66 St. Vincent Street, Glasgow, who sent it to James Watt junior.

The document itself has numerous alterations of wording by James Watt junior which have been removed by us. Watt junior's observations, however, when they deal with matters of substance, must be preserved. He suggests that the first engine of Watt's design made at Carron had an 18-inch and not a 9-inch cylinder. He questions the 'some months' delay after the idea of the condenser had come to Watt before he put it to the test of experiment, and suggests a comparison with Robison's account and Watt's own comment thereon.

At the point where the Harts' account refers to the difficulties with the expansive engines there is a note in the margin:

'The present Mr. James Watt does not believe that rotative engines for factory purposes were at the period alluded to made or intended to work expansively. 3 May 1845. The expansive principle was applied solely to pumping engines. The first double rotative engine in which the expansive principle was used was that of Mr. Lee of Manchester at a much later period. The engines applied to Mines certainly worked expansively, and in one of *them,* such a case as is here stated may have happened, but it undoubtedly was of most rare occurrence.'

Two short paragraphs – the one beginning 'Indeed the greatest annoyance' and the one beginning 'This and a number of other ideas' – have been crossed through by James Watt junior.

There is a marginal note about the slide valve in which James Watt junior says: 'This accords with Mr. James Watt's recollection.' But the following paragraph is corrected in the margin by a note: 'This is all a mistake as the nature of the metal of the nozzles and cylinder never was altered': and then below by a further marginal note: 'This is not so. They began casting for themselves in consequence of Mr. John Wilkinson having stopped the execution of their orders to him.' James Watt junior has also struck out, for the same reason, the words in the text referring to other foundries 'in Birmingham'.

Memoranda of James Watt Esquire, from the Messrs. Harts.

The last year Mr. W[att] was in Scotland the Messrs. H[art], then young men, were invited to pass the evening with him, at the house of his sister in law, Miss MacGrigor Cochrane Street Glasgow.

The conversation turned upon Kinniell and Mr. W[att]'s connexion there with Dr. Roebuck.

Messrs. Hart being natives of Borestoness and well acquainted with the localities, enquired if Mr. Watt had seen the large engine, at the Coaliery commonly called the School Yard Engine. The query appeared to revive old associations. he replied 'Aye lads long before you were born' – smiled, and in a familiar manner, proceeded to inform them that after he made the discovery of the separate condenser, and became connected with Dr. R[oebuck] – to make myself practically acquainted with the working of the Steam Engine, which I before only knew in theory, I took charge of that engine. The engine was one of the largest at that time erected in Scotland being 5 feet 3 inches diameter of Cylinder. I had a suit of pit cloaths made, in which I went down the shaft and like any of the other workmen changed the buckets of the pumps, and overhauled the engine, and continued to do so, till I was completely master of the practical workings of it.

Mr. Hart then enquired if a small engine erected on another of the Coal Pits belonging to the same working, commonly called Taylors pit, was one of his own improved condensing engines. Mr. W[att] replied it was erected by him, but on the Newcomen principle and the reason of its being so erected was that the other the large engine, was overburthened with water, and as the coal workings in Taylors pit, were much higher than the workings of the large engine, he proposed to prevent the water falling into the lower level, to make a small engine to drain the pit by itself, and thereby relieve the larger engine. After this engine was finished and he was satisfied that he practically understood the construction and working of the Newcomen engine Dr. R[oebuck] and I proposed to erect one of my own with the separate condenser of which previously there had only been working models, and for this purpose we ordered a small nine inch cylinder, with air pump and condenser from the Carron Works. The castings were brought to Kinniell as we intended to erect the engine upon the well immediately behind the Kitchen, intending to supply Kinniell House with water. A morning soon after the Castings were laid down Dr. R[oebuck] came into my room, and informed me that his affairs were in disorder – that I was a young man, and he did not wish to involve me in his embarassments, and therefore he would write to an old schoolfellow of his Mr. Bolton of Birmingham who he was confident would take his share of the Patent. That the castings laid down at Kinniell were not used by him, and consequently so long as he was with Dr. R[oebuck] he had not completed an engine on his own principle.

Mr. Hart then enquired if he had a work shop in the University when he was repairing Professor Andersons Steam Engine and made the discovery of the separate condenser – He replied I was too poor and known to few persons at that time – I had then a small work shop in the back building first entry to the North of the Beef Market in King Street.

Mr. Hart being curious to hear every Minutiae [sic] of Mr. Watts discovery remarked that he might perhaps recollect how the first idea of it came into his mind. He replied, 'O Yes perfectly' – it was one Sunday afternoon I had gone to take a walk in the Green of Glasgow, when about half way between the Herds House, which stood near the present site of Nelsons Monument, and Arns Well; my thoughts having been naturally turned to the experiments I had been engaged in for saving heat in the Cylinder, at *that part of the road*, the idea occurred to me that as steam was an elastic vapour it would expand and rush into a previously exhausted space and that if I were to produce a vacuum in a separate vessel and open a communication between the steam in the Cylinder and the exhausted vessel, that such would be the consequence. However it occupied my mind so little, although confident of its success, that being engaged with other more pressing employments at the time, it was some months before I could put it to the test of experiment and it was in that little work shop in King street that I made the original model the results from which confirmed the correctness of my anticipation.

Mr. Hart afterwards enquired if Mr. Watt had heard of the Air Engine invented by the Rev. Mr. Stirling Kilmarnock – Mr. W[att] said no – that he had not been in the habit for some years past of perusing the Scientific Journals, as his eye sight had failed, by intense inspection of minute objects. Mr. Hart then made a pencil sketch of Mr. Stirlings Engine and explained its working – Mr. W[att] enquired what advantage Mr. S[tirling] expected to derive from the use of air instead of water – Mr. H[art] stated he understood that Mr. S[tirling] thought it probable it might be of advantage for the purposes of locomotion, as there would be no water to carry; and also in situations where water could not be easily procured – Another advantage aimed at was economizing of fuel by arresting the heat and making it available a second time – Mr. Watt remarked, I am not fond of offering observations on the inventions of other persons, but when Mr. Stirling has got the engine mounted on wheels, and travelling along the road, I will then be able to give an opinion of it, but Mr. S[tirling] will discover that he has many difficultys to overcome, before he gets it into general use, as I experienced in my own invention. When I began first to construct my engine I found the workmen or Engineers accustomed to the erection of former engines so opinion-ated and obstinate that I had to discontinue employing them, and not only to form my own Engines but also my own Engineers – There was another difficulty I had to encounter in getting them generally introduced – After the engine was constructed, and left to the charge of the person who was to work it, who was usually one of the most intelligent men of the establishment it was made for, and who always was present and assisted when our Engineer put an Engine up, and whose duty, was also to instruct him in the use of all the parts – It never less frequently happened that these men when cleaning part of the engine put it out of working order, especially in many of my first Engines, in which I introduced the expan-sion principle.

Mr. Hart then enquired how the expansive principle was applied – Mr. Watt said suppose the engine was to be a six horse power then the Cylinder was the same as for an 18 horse; but the steam was shut off before the piston had descended through one third of the stroke; and the expansion of the steam in the Cylinder acting against the vacuum in the condenser, was allowed to do the rest – As the Boiler and all the other parts of the engine were of the same size or proportions for one of six horse power therefore if the person in charge of the Engine happened in cleaning to shift the adjusting screws of the regulating valve, so that it was held longer open than one third of the stroke, from the boiler being calculated to supply 1/3d of the Cylinder with steam, it quickly became exhausted and the Engine always stopped after making a few revolutions. When this occurred an express was instantly sent off to Soho informing us that their Engine had gone wrong, and that their factory was at a stand, and the people all off work. We in our part had to write off to the nearest of our men to leave the job they were at, and post over to their work as their Engine had gone wrong, and their factory was standing – On their arrival they found the Engine had stopped solely because the regulating Valve had been altered, and did not shut at the proper time – As this blunder occurred several times from the ignorance of the Engine Keepers, when giving what they termed a thorough cleaning to the Engine, and as it was attended with a serious loss to the Manufacturers, and also gave our Engines a bad characters with the public; under these circumstances we resolved rather to forgoe the saving obtained from using the steam expansively, and by making our Engine less complicated it would be less liable to get out of order, when left to the management of such men.

Indeed the greatest annoyance the company had arose from *this* circumstance, as the blunders of the Engine Keepers were attributed to defects of the Engines; which to a certain extent made them less popular with the public and consequently less in demand.

To give some idea of the loss occasioned by one of these stoppages in the Cornwall Mines, the proprietors stated to Mr. Watt that it cost them about £70 an hour before they could get the water out, and the men set to work again. On Mr. Hart expressing astonish-ment at so large a sum, Mr. Watt observed that these mines were not like coal workings, in which the growth water for a short period would hardly be known, but the Cornish Mines were more like a series of rat holes – from following the ramifications of the veins, and therefore a small quantity of Water in them was quite sufficient to stop a great number of Miners from their work.

This and a number of other ideas that occurred to him for the purpose of economizing fuel he had to abandon, solely for the sake of keeping the Engines as simple as possible, as the saving of a few tons of Coal in the year was nothing compaired to the loss occassioned by some of these stoppages.

He then remembered that when Mr. Murdock introduced the slide valve he set his face against it, because he considered it would occasion a loss of Steam, as they never could make it so perfect, as the puppet Valve, and that its action would soon render it untight from the metals adhering together, whereas the puppet valve had no rubbing service to become leaky; However he observed as Mr. Murdock and my son were fond of the idea, and determined to try it, I gave in, and now I am satisfied it is an improvement after all, as although it may not be so tight as the puppet still it has rendered the Engine much simpler; and there are now fewer parts to go wrong.

He also mentioned another difficulty he met with, in the leaking of the joints between the nozzels, and the Cylinder; arising from the unequal expansion and contraction of the Metal. As to procure good Castings of the nozzels they had to use soft iron and to obtain a durable Cylinder with a fine hard surface, the harder kind had to be resorted to. These not having the same rate of expansion soon rendered the joint untight from the frequent heating and cooling. To obviate this they tried many compounds until they got one that would make a clean cast, and expand and contract in nearly the same ratio as the Cylinder. This was the reason they began casting for themselves, because the castings they formerly got from the other foundrys in Birmingham did not answer for this purpose. Mr. Hart also mentioned that he and his brother had frequent conversations with Mr. Watt when he visited Glasgow, on Optics and other subjects, but that the preceding statement was the principal part of the Conversation so far as he could recollect, the last evening he passed with him.

Two or three years after this conversation with Mr. Watt Mr. Hart being at Borinstonness, met two old men, who he knew were workmen with Dr. Roebuck at the time Mr. Watt was there – He enquired if they recollected Mr. Watt who was with Dr. R[oebuck] about the work – They replied that they did perfectly – Mr. H[art] then enquired if they knew what had become of him – They said no, they never heard of him after he left Dr. Roebuck, but that he was a fine quiet young man – That the person Mr. Watt succeeded was an oppressive tyrannical man, and consequently they were all fond of Mr. W[att] who used to put on pit Cloaths and go down and change the buckets of the pumps with them as a common man – That Mr. W[att] resided at Kinniell House, with Dr. R[oebuck] and generally came down in the morning, arm in arm, with John Roebuck the Drs. son.

In continuation of the jottings of the conversation with Mr. Watt it occurred to Mr. Hart that a few observations on the actual state of the steam engines at that period in the Collieries of Dr. Roebuck at Borrowstoness would be of importance to shew how Mr. W[att] who was bred a Mathematical instrument maker, became so intimately acquainted with the details of the Steam Engine, as to engage at once in the erection of such ponderous Machinery – There were three Steam Engines going at Boness at that period, the one which stood about 100 Yards to the West of the Parish Church was old, being the second Steam engine errected in Scotland, and having the boiler placed beneath the cylinder and also the old hand-gearing similar to the engine described in Desaguliars Work – The Weston Engine, which stood on the site of Mr. Jannings Distillery was erected about the period of the rebellion or shortly after – The Boilers were placed on the outside of the building, which at that period was considered a great improvement, both these engines had Cylinders of about 4 feet diameter. The school yard engine was erected only a few years prior to Mr. Watts connection with Dr. Roebuck and therefore might be considered as put up upon the most approved principles and besides was one of the largest in the country. Here Mr. Watt had an opportunity of inspecting the Steam Engine in all its stages, of taking measures and making drawings etc., and from his going down the pit and inspecting the pumps he could take the thickness of the metal, the size and number of the bolts, and the manner of jointing them, and also the size of the Cisterens at the different lifts, as well as the scantling of the timber by which the pumps are suspended – All this would be an easy process to an ingenious young man, and would afford him all the data necessary for making out the specification even for the Cornwall Engine – Mr. Watt having made himself Master of all this, it would be an easy matter for him to erect the smaller one at Taylors Pit, which having succeeded in doing, he then very properly proceeded to try one of his own – By errecting it at Kinniell House he would have had an opportunity quietly of ascertaining its capabilities before giving it out to the public – however Dr. Roebucks misfortunes put a stop to the erection of the Engine and closed Mr. Watts career at Boness.

5. Extracts from letters between Watt, Roebuck and others, 1765–1769.
   Doldowlod.

The course of the prolonged and often frustrating experiments which Watt carried out with Roebuck's assistance at Kinneil has already been documented by Muirhead in his extensive reproduction of the Watt–Roebuck correspondence. We, therefore, have merely selected a few samples from the letters between Watt, Roebuck, and others in the years 1765–9 (*i.e.* between Watt's first inspired ideas and his patent of 1769), to illustrate, firstly, the thoroughly scientific experimentation which Watt continued to carry out in developing his engine, in preparation for its first large-scale trials; then the preparatory steps towards securing a patent, with supporting specification; and also the first hint of trouble from others (notably Joseph Hateley) who claimed that Watt's ideas were not original. It was, in fact, mainly from fears of his invention being forestalled or pirated that Watt was hurried into getting a patent, despite Roebuck's misgivings, before he had successfully developed the engine, and hence, partly, as we shall see, that his specification only outlined the basic principles of his improvements instead of providing technical details of a fully developed engine – with the ironic result of many future piracies.

(i) JAMES WATT TO JAMES LIND,
29 April 1765, Doldowlod.

'. . . I have now almost a Certainty of the facturum of the fire engine having determined the following particulars of the Quantity of Steam produced—the ultimatum of the Lever engine the Quantity of Steam destroyed by the Cold of its Cylinder the Quantity destroyed in mine and if there is not some devil in the hedge mine ought to raise Water to 44 feet with the same Quantity [of] steam that theirs does to 32 (supposing my Cylinder as thick as theirs) which I think I can demonstrate. I can now make a Cylinder of 2 foot diameter and 3 foot high only a 40th of an Inch thick and strong enough to resist the Atmosphere – *sed tace* in short I can think of nothing else but this Machine. I hope to have the decisive tryal before I see you. . . .'

(ii) JAMES WATT TO DR. JOHN ROEBUCK,
9 September 1765, Doldowlod.

'Dear Sir
  'The Circumstances of the experiment mentioned in my Last were as follows. The Still being made as we agreed, had about a pint of Watter put into it and was close corkt. The Balneum or outer vessel was filled 3 Inch deep with Water. it was then put on the fire and made to boil. the hole at bottom of the Receiver being left open the steam pushed out the air at it. then that hole being close corkt the balneum was filled with Cold water which lowered its heat to that in which watter boils in vacuo, the Receiver was also Imersed in Water at 55°. it was suffered to distill till the water in the Refrigeratory had gained 21°: or was at 76°: the cork at a. was then taken out the air entered with violence, the Refrigeratory being taken away the Cork at b. was also taken out. there issued a measure of water of which the Refrigeratory contained 58. Consequently one measure of water had heated 58 measures 21° so that it must have contained 1218° of heat to which being added the heat it still retained

$$\begin{array}{r} 21 \\ \hline 58 \\ 116 \\ \hline 1218 \\ 76 \\ \hline 1294 \end{array}$$

viz 76° the total will be 1294. the heat of steam in air is only 1028 so here would be an odds of more than $\frac{1}{4}$ of the heat lost by Distilling in Vacuo. Care was taken to prevent the Refrigeratory being heated by other things and Repetitions of the experiment agreed very nearly. any Objections that occur shall answer at meeting when we shall make the Experiment in Company. from this Experiment and some other facts I conclude that in proportion as the sensible heat of steam Increases its latent diminishes so that in the steam engine working with pressures above 15 lb must be more advantageous than below it. for not only the latent heat is diminished but the steam is considerably expanded by the sensible heat which is easily added.

'In the engine Model you saw, I observed that when the steam was just of the heat of boiling or a little above it that it lifted the weight on the Condensing Cock being opened yett on that Cock being shut tho the steam Cock was not opened the piston and weight descended but on using steam a degree or two hotter than boiling tho both Cocks were shut the piston and weight remained suspended – In the first Case the Cylinder having at first Condensed a Quantity of steam was wett on the Inside. on the vacuum being produced hot water was converted into Steam which along with the rest went into the Condenser but that exit being shut the steam still continuing to be produced allowed the piston to descend. in the second case the cylinder being hotter than boiling and filled with an Expended steam (more easily condensible than the other) on the condensing cock being opened the wett sides of the Cylinder would immediately boil violently and before the piston could ascend become entirely dry and it is probable that they would not be cooled much below 212. Consequently they would not deprive the next steam that entered of any of its Latent heat tho they might a little lessen its sensible. Consequently the sides of the Cylinder would still remain Dry. I am going on with the Modell of the Machine as fast as possible and hope to have it finished in another Week.

I remain etc.

WATT

The experiments in this letter seem to have been made on the first model of the improved engine. JW 1808.'

(iii) DR. JOHN ROEBUCK TO JAMES WATT,
13 September 1765, Doldowlod.

'. . . The Fact which you relate is so paradoxical that if less accurate observers of Nature than Yourself and Dr. Black had not been witnesses of it I should scarce have given credit to it. If I rightly understand the process it amounts to this that a certain quantity of water converted into Steam in vacuo will heat a given quantity (of Water) to a greater degree than the same Quantity of Water converted into Steam in Air even though the Steam in Air appears by the Thermometer of a greater heat then the Steam in Vacuo. I shall be glad to try the Experiment with you and make our observations on it afterwards. The Experiment is favourable for Your plan of an Engine and its failing of the other purpose is of less moment. . . .

'. . . I am glad to hear you are in such forwardness to make your tryal. It is certainly best to push it forward with all speed whether you pursue it as a Philosopher or as a Man of Business.'

(iv) G. JARDINE TO JAMES WATT,
No date, 1768, Doldowlod.

'. . . The Doctor has been with you at Glasgow and we have since had some occasional conversations – I wanted to find, without any direct enquiry, if he had in any respect consented to the proposal from the South – but understand, That the more he is convinced of the practibility of the Scheme – The keener he is of carrying it to practice yourselves for your mutual advantage; and I am perfectly convinced, that though he may mistake particulars, it is one of his strongest motives, that you may not only have the honour but the advantage of your own ingenuity. I found that he has this at present so strongly before him, that it would have been vain and perhaps impudent for me to have suggested any thing to the contrary. And therefore my opinion is, James, That you will find it necessary on account of your intimate connection, to fall in with his sentiments and contribute as speedily and as vigorously as you can to bring it to some conclusion. . . . The Dr. proposes not only in the present but also in future undertakings, to have every thing that requires great Nicety and exactness, to be made by the best workmen, and as far as possible to follow the method of Watchmakers, who have little more to do, than to adjust and combine materials – '

(v) DR. JOHN ROEBUCK TO JAMES WATT,
30 October 1768, Doldowlod.

'. . . You are letting the most active part of your life insensibly glide away. A Day a Moment ought not to be lost. And you should not suffer your Thoughts to be diverted by any other object or even improvement of this but only the speediest and most effectual manner of executing one of a proper Size according to Your present Ideas. . . .'

(vi) DR. JOHN ROEBUCK TO JAMES WATT,
9 November 1768, Doldowlod.

'I am every day expecting to see You and Mrs. Watt here. Though I am as sollicitous as Your self that every trial should be made of the small Engine which can be speedily executed particularly the alteration of the Condenser. But then I would not lose a single moment in compleating what was resolved on. I am not so very desirous of hastening the Patent I am afraid we shall be obliged to specify before it is in our interest.'

'The above was forgotten to be sent last night since which time I have received Yours of the 9th. You may bring your various attempts of the specification along with You but we must not think of sending to London for [?] some time. The whole plan of the larger Engine must be fixed when You are here.'

(vii) DR. JOHN ROEBUCK TO JAMES WATT,
16 December 1768, Doldowlod.

'. . . As to the £60 which you have received on account of Fall at Grenock You may if you judge proper remit it to London though I really would not hurry the takeing out of the Patent unless you are afraid least some one should betray You.'

(viii) DR. JOHN ROEBUCK TO JAMES WATT,
22 December 1768, Doldowlod.

'I should be sorry to risk the property of the Engine wherefore by the first Post write to your Friend to take out the Patent. I can spare the money without inconveniency.'

(ix) DR. JOHN ROEBUCK TO JAMES WATT,
19 May 1769, Doldowlod.

[Has closed accounts with Hateley and paid him £50.]
'I hear that he has already sent for a Coppy of Your Specification and that he charges me with revealing his secrets to you. I am very glad however that he cannot charge me with revealing your secrets to himself – '

6. Draft steam-engine specifications

The preparation of the specification for his improvements in the steam engine caused Watt a great deal of mental effort, as shown by the surviving drafts among the Doldowlod papers. Some of these are full of alterations and additions, and several are incomplete. We have selected what appears to be one of the earliest drafts and also a later, more complete version. In the earlier draft, Watt went rather more fully into the defects of the Newcomen engine and the principles on which his improvements were based; his technical description was confined to the 'steam wheel' or rotative engine. In the later draft, he dealt more briefly with the existing defects and his improving principles, but now provided technical details not only about the 'steam wheel', but also about the reciprocating engine, including boiler, cylinder, piston, condenser, valves, and air-pump. He was obviously uncertain as to where he should lay most emphasis, but was tending towards more technical description. Subsequently, however, as we shall see, he was advised to the contrary by Boulton and Small and finally adopted their advice in his printed specification. But in abandoning technical description for principles he was to lay himself open to repeated attacks for inadequacy of specification and consequent efforts to evade or overthrow his patent.

In these drafts Watt refers to technical drawings and he left spaces for insertion of letters relating to engine parts to be shown in the drawings (these spaces we have indicated here by asterisks); but we have not been able to find the relevant drawings. We have, however, reproduced another drawing (Plate 6) made at this time, though Watt did not eventually provide any drawings with his enrolled specification.

(i) EARLY DRAFT STEAM-ENGINE SPECIFICATION BY JAMES WATT, PREVIOUS TO THE 1769 PATENT.
Specifications bundle, MS. No. 5, no date, Doldowlod.

I have found Common steam engines are Imperfect 1st because the Cylinder having been Cooled by the Injection in the preceding stroke Condenses a Considerable Quantity of Steam besides what is necessary to fill it.
2dly because the Vacuum is Imperfect without the Cylinder and all the water in it be cooled down to 90° of Fahrenheits Thermometer. If this was done the Vacuum could be made perfect but the Condensation of Steam the next stroke would much Over ballance the power gained, therefore experience has taught Men that an engine works most advantageously when loaded about half the pressure of the atmosphere.
3dly they have not hitherto been applyed to any Circular Motion without the Intervention of Wheels and Ratches or some similar method of reducing the reciprocating Motion of the piston to a Circular one all of which Contrivances have been attended with Great friction and loss of power.
4thly The present way of making the piston of the Cylinder tight with old ropes etc. is the cause of great friction.

All of these Imperfections are the cause of *much needless Consumption of Steam* and Consequently fuel which I propose to lessen by the following Contrivances.
1stly I propose that the Cylinder by which I mean the principal working Vessel of the engine (whatever its form is) shall never while working be colder than the heat in which water boils or 212° of Fahrenheits Thermometer this will consequently condense no steam.
2dly I propose to exhaust the Cylinder by opening a communication into a Vessel or Vessels that have a part of them emptied of air and water and are always kept colder than 90° of Fahrenheits Thermometer (these vessels or Vessel I call the Condenser) the Steam is here suddenly converted into water by the Coldness of the Condenser, by passing thro cold water in its passage to it, or by any other means by which it can be brought into Contact with Cold bodies in that place: the first Quantity of Steam being thus Condensed the steam remaining in the Cylinder will by its elasticity continue to push more into the Condenser which will also be conden[s]ed untill the Cylinder is perfectly exhausted tho not cooled because I suffer no Cold water to enter it. The water of which the steam was composed and

a small Quantity of air that enters along with it are forced out at valves in the upper part of the Condenser by means of a small pump or pumps wrought by the engine or otherwise. 3dly I propose to Lessen the friction of the piston by two Methods one is by using pasteboard soaked in lintseed or Rape seed Oill and dryed and applyed in the same way as the Leather is commonly fixed on the bucket of a sucking pump this is both much lighter than the Common engine piston and has Less friction the second is by surrounding the piston with Quicksilver or any metal that will be fluid in the heat of boiling water.

4thly I propose to apply the fire engine to make bodies turn round; in the following manner. [There follows a detailed technical description of the 'steam wheel', but as this is very closely similar to that given at the end of the following document, we have not included it here.]

In cases where Cold water cannot be Obtained for the Condenser I use steam much stronger than that of boiling water and made to act only by its superiority of pressure to that of the Atmosphere in these Cases Instead of opening a Communication with the Condenser one is opened to the external air this is applicable to either the reciprocating or Circulating mode of my engine.

[Watt then wrote a conclusion, summarising his improvements, followed by a revised version, but as the latter was a clearer, more concise rewording of the former, and since Watt himself considered it 'Stronger than the other and better Calculated for catching the attention', we have only reproduced this revised conclusion.]

To Conclude, The Excellency and peculiarity of this Method principaly consists In Keeping the steam Vessel or Cylinder always as hot as the steam, while the Engine is at work, 2dly In Condensing the Steam in a vessel distinct from the Steam vessel but communicating with it during the time of condensation, by these two I prevent the Alternate heating and cooling of the Steam vessel which destroys much steam in Common fire engines and by the Constant cold of the Condenser I condense the steam more perfectly and acquire a Greater power. 3dly In Employing the steam to press on the piston and thereby being enabled to increase the power beyond the pressure of the Atmosphere. 4thly It consists in making the steam vessel Itself turn round on an axis (when the case requires it) by the force of Steam Acting against a weight contained within it and thereby preventing the friction and Complex Machinery Occasioned by deriving Circulating from reciprocating motion.

(ii) DRAFT STEAM-ENGINE SPECIFICATION BY JAMES WATT, PRIOR TO THE 1769 PATENT.
Specifications bundle, MS., no number or date, Doldowlod.

I find that in a common fire engine, the atmosphere will not press on the piston with its whole weight, unless the Cylinder be cooled to 90° of Fahrenheits Thermometer because water boils much more easily in Vacuo than in the open air.

I find also that the Cylinder being cooled by the Injection reduces while it is filling a great quantity of the Steam into Water –

Experience has shown that it is disadvantageous to make the power exerted by this engine greater than one half the pressure of the Atmosphere on its Piston –

It is well known that the Application of Reciprocating fire Engines to turn milns or other circular motions has always been attended with great loss of power – Every method of increasing the power of the engine, or of diminishing the quantity of steam now requisite, also lessens c[eteris] p[aribus] the consumption of fuel –

My method of increasing the power is,

By condensing the steam in a place distinct from the Cylinder, but occasionally communicating with it, previously exhausted of air and which may be made as cold as is necessary for the total condensation of the Steam without cooling the Cylinder in the least –

I also lessen the consumption of Steam by keeping the Cylinder as near as possible as hot as the Steam issuing from the boiler, and by never allowing water to enter it – The Cylinder thus remaining nearly as hot as the Steam issuing from the boiler and no water being ever allowed to enter it.

As in this method no water ever enters the Cylinder, it will remain nearly as hot as the Steam issuing from the boiler, and consequently never can condense any Steam – And as the place where the Steam is condensed (which I call the Condenser) may be made even colder than 90° the exhaustion will be perfect, and the atmosphere made to press on the piston of the Cylinder with nearly the whole of its weight –

I also increase the power of the engine by employing the Steam issuing from the boiler to press upon the Piston of the Cylinder instead of the Atmosphere –

My Method of applying the fire engine to move milns etc. is by using a hollow wheel instead of the Cylinder, which is made to turn round *on its axis* by the force of Steam acting against a weight contained within it –

This is the substance of my method of lessening the consumption of Steam and fuel in fire Engines, this is of my Invention and is totaly different from, and far superior to any thing that has ever been proposed for that purpose – What I have said is sufficient to enable the skilful mechanic to construct fire engines with the proposed properties – I have however annexed drawings with explanations to show how this may be reduced to practice and to describe some things which tho' less important are however necessary in the execution –

The Cylinder * is covered with a lid * thro the center of which the Piston stalk * slides in a Collar of Pasteboard or Oakum steam tight. This end of the Cylinder communicates with the boiler by a Pipe joining the steam pipe at * and always allowing free passage for the steam to press on the Piston when wanted. At the lower end of the Cylinder is fixed a Regulating Valve capable of making a communication either between the Boiler and the Cylinder or between the Cylinder and the Condenser *. The Condenser is not confined to any certain form tho the more surface that can be given to it the better. it shoud be very thin but must be air tight and strong enough to resist the pressure of the Atmosphere   a part of it must be emptied both of air and water – This may be done by a Pump * in the following manner. The Piston * of this Pump being at the bottom the Condenser is to be filled quite full of water which is prevented from going into the Cylinder by a Valve *. Then the regulating Valve being open to steam and shut to the Condenser the Piston * is to be raised up the air contained in the Pipe * will then come into the Condenser thro the Valve * and the Piston * being allowed to descend will expel that air at the small Valve *. This operation repeated will leave the Condenser and pipe * perfectly free from air.

The Piston * being raised up will leave the upper part of the Condenser free both of air and water. The Cylinder * is now to be filled with steam and the air and water let out at a Cock * which as soon as the Cylinder is thoroughly heated is to be shut by hand.

The Regulating Valve * is now to be shut to the boiler and opened to the Condenser; a part of the steam contained in the Cylinder will immediately rush into the condenser which being immersed in cold water or having a stream of cold water running over it will instantly reduce it to Water more Steam will come from the Cylinder to be condensed in the same manner till the Cylinder is perfectly exhausted. The Piston * will then descend and raise the load required. The regulating valve will then be shut to the condenser and open to the boiler and the Cylinder filled with Steam as before and the Piston of the Condenser raised ready for the next stroke – NB As no water can in this engine be made use of to keep the Piston of the Cylinder tight its place must be supplied by oil or melted tallow – By the Cylinder I always mean the principal working Vessel of the machine whatever its form may be –

Description of the Circular moving Fire Engine. A Wheel is to be made with a hollow channel (*close on all sides*) round its circumference – The size of this wheel and the channel round it are regulated by the power required – In this channel at proper Intervals are placed a number of valves not fewer than three which will allow any body to pass round the wheel in one direction but will allow nothing to go the contrary way. The axis * of this wheel is hollow for some space at each end and terminates in Cones * which are fitted into hollow Cones *. The one * communicating with the boiler * and the other * with the Condenser – From the hollow part of the axis on each side of the wheel proceed as many Tubes as there are Valves in the Channel round the wheel   the other end of each of these tubes is inserted into this Channel near the Valves and is there shut by a regulating Valve *. A quantity of Quicksilver or oil or any other fluid which does not boil so easily as water must be poured into the Channel * and will naturaly run to the lowermost side. All communication betwixt the Channel and the external air being now cut of[f] and the Channel exhausted of air a Regulating Valve * is to be opened to allow the Steam to enter betwixt a great Valve * and the surface of the fluid this Valve being shut the steam will press equaly on it and on the surface of the fluid in the Channel the fluid will retreat from the Valve and rise on the other side of the wheel till the excess of its weight on that side is equal to the power required as the wheel will turn round in the Opposite direction more Steam will enter and support the fluid in the Channel to the height to which it was raised, another great Valve * will emerge from the fluid and be shut, another regulating Valve * is at the same time opened the first regulating Valve must now be shut the corresponding great Valve * which will now be in the position * and a regulating Valve communicating with the Condenser opened the Steam contained in the space * will rush into the condenser and leave that part of the Channel exhausted – The wheel will continue to go round and the same operation will be repeated on each of the Valves in Succession – So that the engine will work uniformly

53

as long as it is supplied with Steam – In this application of my method the friction is much diminished by using a fluid instead of a Piston and no power is spent in Overcoming the resistance of matter to motion as is the case in all reciprocating fire engines – In order to diminish the quantity of Quicksilver necessary I employ a lump of iron fitted as well to the Channel as may be yet so as to move easily in it and the interstices are filled with quicksilver this will have little more friction than if quicksilver alone had been made use of and will be abundantly cheaper.

NB In cases where cold water cannot be obtained for the Condenser I use Steam much stronger than that of boiling water and made to act only by its superiority of Pressure to that of the atmosphere in these cases instead of opening a communication with the Condenser one is opened to the external air this is applicable to either Reciprocating or Circulating Engines.

7. Dr. William Small to James Watt,
5 February 1769, Doldowlod.

Roebuck's business difficulties eventually obliged him to negotiate for the sale of a part or even the whole of his share in Watt's engine to Matthew Boulton, of the Soho works, near Birmingham, in conjunction with Dr. William Small. These negotiations began even before Watt had acquired his patent (see above, p. 48, for reference to 'the proposal from the South'). This letter gives the reactions of Boulton and Small to Watt's draft specification. They, as previously mentioned, were strongly of opinion that technical details should not be included and that only the general principles of Watt's invention should be explained, in order to guard against piracies.

'Dear Sir,

'Mr. Boulton and I have considered your paper, and think you should neither give drawings nor descriptions of any particular machinery, (if such omissions would be allowed at the office) but specify in the clearest manner you can, that you have discovered some principles, and thought of new applications of others, by means of both which joined together, you intend to construct steam engines of much greater powers, and applicable to a much greater number of useful purposes than any which hitherto have been constructed, that to effect each particular purpose you design to employ particular machinery, every species of which may be ranged in two classes. One class for producing reciprocal motions, and another for producing motions round axes.

'As to your principles, we think they should be enunciated (to use an hard word) as

3 Model of the Newcomen engine by J. Sisson. *By courtesy of Glasgow University*

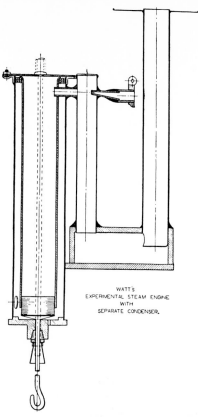

WATT'S
EXPERIMENTAL STEAM ENGINE
WITH
SEPARATE CONDENSER.

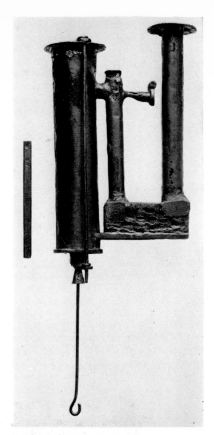

4 Sectional drawing of Watt's separate condenser, from H. W. Dickinson, *James Watt, Craftsman and Engineer* (NB This drawing shows the model conjecturally restored and not in its present unworkable condition)

5 Watt's separate condenser. *Crown copyright. Science Museum, London*

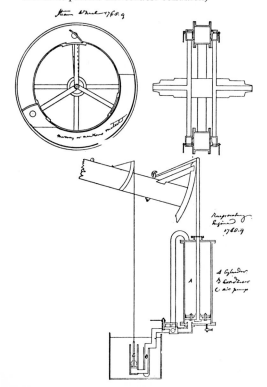

6 Drawings in preparation of the 1769 Patent specification. *By courtesy of Major David Gibson-Watt, M.C., M.P.*

generally as possible, to secure you as effectually against piracy as the nature of your invention will allow. You might declare in some such manner as the following.

'First. You intend that the vessels in which the power of steam will be employed to work such engines as you may construct, shall be heated before the working of the engines shall begin, at least as hot as the steam to be conveyed into the vessels; and that this heat of the vessels shall be rendered equable, whilst the engines work, by suffering them to be entered or touched in that time by no substance colder than the steam they are designed to receive, by covering them with materials which allow bodies so covered to cool very slowly, and by proper applications of heated bodies when they may be wanted. The vessels mentioned in this paragraph you call STEAM-VESSELS.

'Secondly. In engines which you may erect to be worked either wholly or partially by condensation of Steam, you intend that the steam shall be condensed in vessels distinct from the steam vessels, tho occasionally communicating with them. These vessels you call condensers; and, whilst the engines may be working, you intend to keep the condensers constantly at least as cold as the air then in the neighbourhood of the engines, by applications of water and other means of cooling heated bodies.

'Thirdly. Whatever air, or other uncondensable elastic vapor, may impede the operations of the engines you intend shall be drawn out by machines, in the manner of pumps; to be worked by the engines themselves.

'Fourthly. You intend that on different occasions the necessary steam shall be produced from different substances, solid or fluid, or partly solid and partly fluid, as may be most convenient; and also that the vessels in which the steam shall be produced (which you call boilers), shall be of different forms on these different occasions.

'Fifthly. In many cases you design to employ steam in producing reciprocal motions in a manner like to that in which portions of the atmosphere are now employed in ordinary reciprocating engines, to wit by pressing at proper times upon pistons of proper structures.

'Sixthly. To produce by means of steam motions round axes, you intend sometimes to employ reciprocating joined to other machines, but more frequently steam vessels of forms fitted to different purposes. These steam vessels will be mounted on axes, and will contain weights, either solid or fluid, or partly solid and partly fluid, which weights, or the centres of their gravity, being constantly whilst the engines work pressed by steam beyond planes perpendicular to the horizon, and in which planes the axes will ly, will cause motions of the steam vessel.

'Seventhly. In these last mentioned engines, in which steam vessels must move round axes, on some occasions I [*sic*] intend to use the condensers described above, but on others to discharge the steam from the steam vessels thro proper outlets into the atmosphere.

'Lastly. To render pistons and other p[arts of] the machinery air and steam tight, instead of water you design to employ paper and paste board prepared with oils, oils themselves, or fats of animals, quicksilver or melted metals.

'February 7th. I have received your letter of the 28th January. Upon the whole an Act of Parliament would be the thing, you obliging yourself to reveal to Commissioners or to the Royal Society every particular proper to be revealed. I am certain that from such a specification as I have written any skilful mechanic may make your engines, altho it wants corrections, and you are certainly not obliged to teach any blockhead in the nation to construct masterly engines. I wish we could any how meet before you specify. A matter of this kind cannot be managed by letters. I should have much to say to you about your machinery. As to your invention for measuring distance I know of nothing similar to it, excepting what I have formerly told you I had recommended in a case in which I was applied to as a mathematical man, and which is not so easy as yours, and another method I thought of for measuring the heighths [*sic*] of mountains which I never told you. I have invented little of late. An American friend of mine has discovered a method of making nitre monstrously cheap, but I do not know the method. He spent lately a week with me, but I did not care to know his secret before he had secured it.

'I am glad the alcali is attended to. I know nothing for which I would more willingly have quited physic, if the invention had been mine. Boulton has written to you. Come to live in England as soon as you can. Till then do not fatigue yourself, and never think, and let Mrs. Watt nurse you mighty well. Do make my compliments to her. Farewell. Captain Kier is in Berkshire. Mrs. Darwin has been long ill.

W. SMALL'

8. Letters Patent, and Specification of Patent.
1769

(i) LETTERS PATENT, DATED JANUARY 5TH, 1769, FOR A METHOD OF LESSENING THE CONSUMPTION OF STEAM AND FUEL IN FIRE ENGINES.

GEORGE THE THIRD, BY THE GRACE OF GOD, &C.
TO ALL TO WHOM these presents shall come, greeting:
WHEREAS JAMES WATT hath by his Petition humbly represented unto us that he hath, after much labour and expense, invented a Method of lessening the Consumption of Steam and Fuel in Fire Engines, which he apprehends will be of great Publick Utility: And in regard he is the first and true Inventor thereof, and the same hath not been practised or used by any other person or persons to his knowledge or belief, he therefore most humbly prayed us to grant unto him, his executors, administrators, and assigns, OUR ROYAL LETTERS PATENT under Our Great Seal of Great Britain, for the sole benefit and advantage of his said Invention within that part of Great Britain called England, our Dominion of Wales, and Town of Berwick upon Tweed, and also in our Colonies and Plantations abroad, for the term of Fourteen years, according to the Statute in that case made and provided: WE, being willing to give encouragement to all Arts and Inventions which may be for the Publick Good, are graciously pleased to condescend to his Request.

KNOW YE, therefore, that WE of our especial grace, certain knowledge and meer motion, Have given and granted, and by these presents for us, our heirs and successors, Do give and grant unto the said JAMES WATT, his executors, administrators, and assigns, OUR especial licence, full power, sole priviledge and authority, that he the said JAMES WATT, his executors, administrators, and assigns, and every of them, by himself and themselves, or by his and their deputy or deputys, servants or agents, or such others as he the said JAMES WATT, his executors, administrators, or assigns, shall at any time agree with, and no others, from time to time and at all times hereafter during the term of years herein expressed, shall and lawfully may make, use, exercise and vend his said Invention within that part of our Kingdom of Great Britain called England, our Dominion of Wales, and Town of Berwick upon Tweed, and also in our Colonies and Plantations abroad, in such manner as to him the said JAMES WATT, his executors, administrators, and assigns, or any of them, shall in their discretions seem meet. AND that he the said JAMES WATT, his executors, administrators, and assigns, shall and lawfully may have and enjoy the whole profit, benefit, commodity, and advantage, from time to time coming, growing, accruing, and arising by reason of his said Invention, for and during the term of years herein mentioned: To have, hold, exercise, and enjoy the said licence, powers, priviledges, and advantages herein before granted or mentioned to be granted unto the said JAMES WATT, his executors, administrators, and assigns, for and during and unto the full end and term of Fourteen years from the date of these presents next and immediately ensuing, and fully to be compleat and ended according to the Statute in such case made and provided. AND to the end that he the said JAMES WATT, his executors, administrators, and assigns, and every of them, may have and enjoy the full benefit and the sole use and exercise of the said Invention, according to our gracious intention herein before declared, – WE do by these presents, for us, our heirs and successors, require and strictly command all and every person and persons, bodies politick and corporate, and all other our subjects whatsoever, of what estate, quality, degree, name, or condition soever they be, within that said part of our Kingdom of Great Britain called England, our Dominion of Wales, and Town of Berwick upon Tweed, and also in all our Colonies and Plantations abroad aforesaid, that neither they nor any of them, at any time during the continuance of the said term of Fourteen years hereby granted, either directly or indirectly do make, use, or put in practice the said Invention, or any part of the same, so attained unto by the said JAMES WATT as aforesaid, nor in any wise counterfeit, imitate, or resemble the same, nor shall make or cause to be made any addition thereunto or subtraction from the same, whereby to pretend himself or themselves the Inventor or Inventors, Devisor or Devisors thereof, without the licence, consent, or agreement of the said JAMES WATT, his executors, administrators, or assigns, in writing under his or their hands and seals first had and obtained in that behalf, upon such pains and penalties as can or may be justly inflicted on such Offenders for their contempt of this Our Royal Command: AND further to be answerable to the said JAMES WATT, his executors, administrators, and assigns, according to Law, for his and their damages thereby occasioned. And moreover WE do by these presents for us, our heirs and successors, will and command all and singular the Justices of the Peace, Mayors, Sheriffs, Bailiffs, Constables, Headboroughs, and all other Officers and Ministers whatsoever of us, our heirs and successors for the time being, that they or any of them do not nor shall at any time hereafter during the said term hereby granted in anywise molest, trouble, or hinder the said JAMES WATT, his executors, administrators, or assigns, or any of them, or his or their deputys, servants, or agents, in or about the due and lawful use or exercise of the aforesaid Invention, or anything relating thereto. PROVIDED ALWAYS, and these our Letters Patent are and shall be upon this Condition, that if at any time during the said term hereby granted it shall be made appear to us, our heirs or successors, or any six or more of our or their Privy Council, that this our Grant is contrary to Law, or prejudicial or inconvenient to our subjects in general, or that the said Invention is not a new Invention as to the publick use and exercise thereof in that said part of our Kingdom of Great Britain called England, our Dominion of Wales, and Town of Berwick upon Tweed, and our Colonies and Plantations abroad aforesaid, or not invented and found out by the said JAMES WATT as aforesaid, – Then upon Signification or Declaration thereof to be made by us, our heirs or successors, under our or their Signet or Privy Seal, or by the Lords and others of our or their Privy Council, or any six or more of them, under their hands, these our Letters Patent shall forthwith cease, determine, and be utterly void to all intents and purposes, anything hereinbefore contained to the contrary thereof in anywise notwithstanding. PROVIDED ALSO, that these our Letters Patent, or anything herein contained, shall not extend or be construed to extend to give priviledge unto the said JAMES WATT, his executors, administrators, or assigns, or any of them, to use or imitate any Invention or Work whatsoever which hath heretofore been found out or invented by any other of our subjects whatsoever, and publickly used or exercised in that said part of our Kingdom of Great Britain called England, our Dominion of Wales, or Town of Berwick upon Tweed, or our Colonies and Plantations abroad aforesaid, unto whom like Letters Patent or priviledges have been already granted for the sole use, exercise, and benefit thereof; it being Our will and pleasure that the said JAMES WATT, his executors, administrators, and assigns, and all and every other person and persons to whom like Letters Patent or priviledges have been already granted as aforesaid, shall distinctly use and practise their several Inventions by them invented and found out, according to the true intent and meaning of the same respective Letters Patent and of these presents; provided likewise nevertheless, and these our Letters Patent are upon this express Condition, that if the said JAMES WATT, his executors or administrators, or any person or persons which shall or may at any time or times hereafter during the continuance of this grant, have or claim any right, title, or interest in Law or in Equity, of, in, or to the power, priviledge, and authority of the sole use and benefit of the said Invention hereby granted, shall make any transfer or assignment, or any pretended transfer or assignment of the said liberty and priviledge, or any share or shares of the benefits or profit thereof, or shall declare any Trust thereof to or for any number of persons exceeding the number of five, or shall open or cause to be opened any book or books for publick subscriptions to be made by any number of persons exceeding the number of five, in order to the raising any sum or sums of money under pretence of carrying on the said liberty or priviledge hereby granted, or shall by him or themselves, or his or their agents or servants, receive any sum or sums of money whatsoever of any number of persons exceeding in the whole the number of five, for such or the like intents or purposes, or shall presume to act as a corporate body, or shall divide the benefit of these our Letters Patent, or the liberty and priviledges hereby by us granted, into any number of shares exceeding the number of five, or shall commit or do, or procure to be committed or done, any act, matter, or thing whatsoever during such time as such person or persons shall have any right or title, either in Law or Equity, in or to the said premises, which will be contrary to the true intent and meaning of a certain Act of Parliament made in the Sixth year of the Reign of our late Royal Great Grandfather, King George the First, intituled 'An Act for the better securing certain powers and priviledges 'intended to be granted by His Majesty by two Charters for Assurance of Ships and Mer-'chandizes at Sea, and for lending Money upon Bottomry, and for restraining several extra-'vagant and unwarrantable practices herein mentioned,' – or in case the said power, privi-ledge, or authority shall at any time hereafter become vested in or in Trust for more than the number of five persons, or their representatives at any one time (reckoning executors or administrators as and for the single person whom they represent as to such interest as they are or shall be intituled to in right of such their testator or intestate), That then, and in any of the said cases, these our Letters Patent, and all liberties and advantages whatsoever here-by granted, shall utterly cease, determine, and become void, anything hereinbefore con-tained to the contrary thereof in anywise notwithstanding. PROVIDED ALSO, that if the said JAMES WATT shall not particularly describe and ascertain the nature of his said Invention, and in what manner the same is to be performed, by an instrument in writing under his hand and seal, and cause the same to be inrolled in our High Court of Chancery within four calendar months next and immediately after the date of these our Letters Patent, that then these our Letters Patent, and all liberties and advantages whatsoever hereby granted, shall utterly cease, determine, and become void, anything hereinbefore contained to the contrary thereof in anywise notwithstanding. And LASTLY, WE do by these presents, or us, our heirs and successors, Grant unto the said JAMES WATT, his executors, adminis-trators, and assigns, that these our Letters Patent, or the Inrolment or Exemplification thereof, shall be in and by all things good, firm, valid, sufficient and effectual in the Law, according to the true intent and meaning thereof, and shall be taken, construed, and ad-judged in the most favourable and beneficial sense for the best advantage of the said JAMES WATT, his executors, administrators, and assigns, as well in all our Courts of Record as elsewhere, and by all and singular the Officers and Ministers whatsoever of us, our heirs and successors, in that part of our said Kingdom of Great Britain called England, our Dominion of Wales, and Town of Berwick upon Tweed, and also in our Colonies and Plantations abroad aforesaid, and amongst all and every the subjects of us, our heirs and successors, whatsoever and wheresoever, notwithstanding the not full and certain describing the nature or quality of the said Invention, or of the materials thereto conducing and belonging. In Witness, &c.

Witness Ourself at Westminster, the fifth day of January, in the ninth year of our reign.
BY WRITT OF PRIVY SEAL.

(ii) SPECIFICATION OF PATENT, JANUARY 5TH, 1769, FOR A NEW METHOD OF LESSENING THE CONSUMPTION OF STEAM AND FUEL IN FIRE ENGINES.

To ALL TO WHOM these presents shall come, I, JAMES WATT, of Glasgow, in Scotland, Merchant, send greeting.

WHEREAS His Most Excellent Majesty King George the Third, by his Letters Patent, under the Great Seal of Great Britain, bearing date the fifth day of January, in the ninth year of his said Majesty's reign, did give and grant unto me, the said JAMES WATT, his special licence, full power, sole privilege and authority, that I, the said JAMES WATT, my executors, administrators, and assigns, should, and lawfully might, during the term of years therein expressed, use, exercise, and vend throughout that part of his Majesty's Kingdom of Great Britain called England, the Dominion of Wales, and Town of Berwick upon Tweed, and also in his Majesty's Colonies and Plantations abroad, my new invented 'METHOD OF 'LESSENING THE CONSUMPTION OF STEAM AND FUEL IN FIRE ENGINES;' in which said recited Letters Patent is contained a Proviso, obliging me, the said JAMES WATT, by writing under my hand and seal, to cause a particular description of the nature of the said invention to be inrolled in his Majesty's High Court of Chancery, within four calendar months after the date of the said recited Letters Patent, as in and by the said Letters Patent and the Statute in that behalf made, relation being thereunto respectively had, may more at large appear:

NOW KNOW YE, that in compliance with the said Proviso, and in pursuance of the said Statute, I, the said JAMES WATT, do hereby declare that the following is a particular description of the nature of my said invention and the manner in which the same is to be performed (that is to say): MY METHOD of lessening the consumption of steam, and consequently fuel, in fire engines, consists of the following principles:

FIRST, that vessel in which the powers of steam are to be employed to work the engine, which is called the *Cylinder* in common fire engines, and which I call the *Steam Vessel*, must, during the whole time the engine is at work, be kept as hot as the steam that enters it; first, by inclosing it in a case of wood, or any other materials that transmit heat slowly; secondly, by surrounding it with steam or other heated bodies; and, thirdly, by suffering neither water nor any other substance colder than the steam to enter or touch it during that time.

SECONDLY, in Engines that are to be worked wholly or partially by condensation of steam, the steam is to be condensed in vessels distinct from the steam vessels or cylinders, although occasionally communicating with them: these vessels I call *Condensers*; and, whilst the engines are working, these condensers ought at least to be kept as cold as the air in the neighbourhood of the engines, by application of water, or other cold bodies.

THIRDLY, whatever air, or other elastic vapour, is not condensed by the cold of the condenser, and may impede the working of the engine, is to be drawn out of the steam vessels or condensers by means of pumps, wrought by the engines themselves, or otherwise.

FOURTHLY, I intend in many cases to employ the expansive force of steam to press on the pistons, or whatever may be used instead of them, in the same manner as the pressure of the atmosphere is now employed in common fire engines: in cases where cold water cannot be had in plenty, the engines may be wrought by this force of steam only, by discharging the steam into the open air after it has done its office.

FIFTHLY, where motions round an axis are required, I make the steam vessels in form of hollow rings, or circular channels, with proper inlets and outlets for the steam, mounted on horizontal axles, like the wheels of a water-mill; within them are placed a number of valves, that suffer any body to go round the channel in one direction only: in these steam vessels are placed weights, so fitted to them as entirely to fill up a part or portion of their channels, yet rendered capable of moving freely in them by the means hereinafter mentioned or specified. When the steam is admitted in these engines, between these weights and the valves, it acts equally on both, so as to raise the weight to one side of the wheel, and by the re-action on the valves successively, to give a circular motion to the wheel, the valves opening in the direction in which the weights are pressed, but not in the contrary: as the steam vessel moves round, it is supplied with steam from the boiler, and that which has performed its office may either be discharged by means of condensers, or into the open air.

SIXTHLY, I intend, in some cases, to apply a degree of cold not capable of reducing the steam to water, but of contracting it considerably, so that the engines shall be worked by the alternate expansion and contraction of the steam.

LASTLY, instead of using water to render the piston, or other parts of the engines, air and steam-tight, I employ oils, wax, resinous bodies, fat of animals, quicksilver, and other metals, in their fluid state.

IN WITNESS whereof I have hereunto set my hand and seal this twenty-fifth day of April, in the year of our Lord one thousand seven hundred and sixty-nine.

JAMES WATT.

Sealed and delivered in the presence of

COLL. WILKIE.
GEO. JARDINE.
JOHN ROEBUCK.

BE IT REMEMBERED, that the said JAMES WATT doth not intend that anything in the Fourth Article shall be understood to extend to any engine where the water to be raised enters the steam vessel itself, or any vessel having an open communication with it.

JAMES WATT.

Witnesses. COLL. WILKIE.
GEO. JARDINE.

INROLLED the twenty-ninth day of April, in the year of our Lord one thousand seven hundred and sixty-nine.

9. Matthew Boulton to James Watt,
7 February 1769, Doldowlod.

Meanwhile Roebuck's negotiations with Boulton and Small were still proceeding. The following letter gives Boulton's impressively ambitious and imaginative response to an offer from Roebuck of a part-share in Watt's engine, covering the counties of Warwick, Stafford, and Derby: he was not interested in making engines for three counties only, but was willing 'to make for all the World', by adopting mass-production methods.

'Dear Watt

'By this time I dare say you have fully concluded that I am a very queer fellow, I haveing never answered your friendly letter of the 20th October nor your last of the 12th December – in troth 'tis a shame, and I ask you ten thousand pardons, I could make many excuses and tell you how much I have been engaged and hurried at London, Bath, Bristol, all over Derbyshire, etc. etc. with sundry other reasons all of which I shall wave untill I have the pleasure of seeing you at Soho.

'I note what you say in respect to your connection with Dr. Roebuck from whom I received a letter dated the 12 December offering me a share of his property in your Engine as far as respects the countys of Warwick, Stafford, and Derby; I am obliged to you and him for thinking of me as a partner in any degree but the plan proposed to me is so very different from that which I had conceived at the time I talked with you upon that subject that I cannot think it a proper one for me to meddle with, as I do not intend turning Engineer. I was excited by two motivs to offer you my assistance which were love of you and love of a money-getting ingenious project. I presum'd that your Engine would require mony, very accurate workmanship, and extensive correspondence, to make it turn out to the best advantage; and that the best means of keeping up the reputation, and doing the invention justice, would be to keep the executive part out of the hands of the multitude of empirical Engineers who from ignorance, want of experience; and want of necessary convenience, would be very liable to produce bad and inaccurate workmanship; all which deficiencies would affect the reputation of the invention. to remedy which and to produce the most profit, my idea was to settle a manufactory near to my own by the side of our Canal, where I would erect all the conveniences necessary for the completion of Engines and from which Manufactory We would serve all the World with Engines of all sizes; by these means and your assistance we could engage and instruct some excellent workmen who (with more excellent tools than would be worth any mans while to procure for one single engine) could execute the invention 20 Per Cent cheaper than it would be otherwise executed, and with as great a difference of accuracy as there is between the Blacksmith and the mathematical instrument maker: it would not be worth my while to make for three Countys only, but I find it very well worth while to make for all the World. What led me to drop the hint I did to you was the possesing an idea that you wanted a midwife to ease you

62

of your burthen, and to introduce your brat into the world which I should not have thought of, if I had known of your preengagement but as I am determined never to embark in any trade that I have not the inspection of myself, and as my engagements here will not permit me to attend any business in Scotland, and as the docters engagements in Scotland will not, I presume, permitt his attendance here, and as I am almost saturated with undertakeings, I think I must conclude to – no you shall draw your conclusion, yet nevertheless let my conclusions be what they will, nothing will alter my inclination for being concerned with you, or for rendering you all the service in my power; and although there seem to be some obstructions to our partnership in the Engine trade yet I live in hope that you or I may hit upon some scheme or other that may associate us in this part of the world which would render it still more agreeable to me than it is, by the acquisition of such a neighbour.

The digester, pipes, Kaolin, Flutes etc. are come to hand a few days ago but have not yet had time to inspect them. What's mine is also Dr. Smalls. I beg youl forward the furnice as soon as 'tis ready it being much wanted both by Dr. Small and myself. I presume the furnice flutes and digester may come to about Ten pounds and therefore I have enclosed a Bill for that sum dated the 28th January (for on that day I began to write to you but was obstructed). so soon as you have sent one furnice I beg you will order another the first being for Dr. Small. I am glad to hear that you have a Modell of your reciprocating [engine] in great forwardness, I being impatient to hear of its success. Query is the wheel Engine invention also included in your transfer to the Dr. or can it be separated from it with propriety. I hope it will not be long before you make this place in your way to London and pray remember when you come always to put up at the Hotel d'amitie sur Handsworth Heath where you will always find a friendly reception from

Dear Sir

Your affectionate humble Servant

MATTHEW BOULTON'

10. James Watt to Dr. John Roebuck, *24 September 1769, Doldowlod.*

This letter, from Watt to Roebuck, reveals the very favourable impression which Boulton's character, business enterprise, and technical skill had made on him, and Watt's typical depreciation of his own capacities. He was obviously very anxious to conclude the negotiations between Roebuck and Boulton, and to team up with the latter, who would be better able to supply the necessary capital and technical resources to make the engine a success.

'. . . . When you are at Birmingham I expect you will try what can be done with Mr. Bolton and hope you will be able to conclude some Bargain which even tho it should appear a little hard for us I would wish you to accept from the following considerations.

'1st from Mr. Boultons own Character as an Ingenious honest and rich man. You know him much better than I do, but the worst I have ever heard of him amounted to his being a projector.

'2dly from the difficulty and Expence there would be of procuring Accurate and honest workmen and providing them in proper Utensils and getting a proper Overseer or Overseers. If to avoid this Inconvenience You were to contract for the work to be done by a master workman you must give up a great share of the profit.

'3dly the success of the engine is yett far from being verified. If Mr. Boulton takes his chance of success from the account I shall write Dr. Small and pays you any adequate share of the money laid out, It lessens your risk and in a greater proportion than I think it will Lessen your profit.

'4thly the Assistance of Mr. Boultons and Dr. Smalls Ingenuity (If the later engage in it) in Improving and perfecting the Machine may be very Considerable and may enable us to gett the better of Difficultys that might otherwise damn it. lastly, consider my uncertain health my Irresolute and Inactive disposition My Inability to Bargain and Struggle for my own with Mankind all which disqualify me for any Great Undertaking.

'on Our side Consider the first Outlay and Interest the patent, the present Engine about £200 (tho there would not be much Loss in Making it into a common Engine). 2 years of my time and the expence of models. the post goes off. Compliments to Mrs. Roebuck. farewell

JAMES WATT'

11. Dr. John Roebuck to Dr. William Small and Matthew Boulton, *26 November 1769, Doldowlod.*

A further illustration of Roebuck's negotiations with Boulton and Small is provided by the following letter.

'Gentlemen   Whereas Mr. James Watt has assigned to me two thirds of the Property of the Patent of the Steam Engine which he took out sometime in the Course of the last Winter I hereby offer you one half of the above two thirds or one third of the whole Patent on condition that you pay to me such a Sum not less than One Thousand Pounds as You, anent the Experiments of the Engine shall be compleated, shall think just and reasonable. And twelve Months from this date You are to take Your final final resolution. I also oblige myself to procure Mr. Watts assent to this agreement – I am Gentlemen Your Obedient Servant – JOHN ROEBUCK'

12. Mrs. Margaret Watt to James Watt,
22 [?] 1769, Doldowlod.

This letter to Watt (at Kinneil) from Margaret, his first wife (in Glasgow), is interesting for the further reference to Joseph Hateley, who was now accusing Roebuck and Watt of stealing his ideas for an improved engine.

'Dear Jamie

'I am very happy to hear you are better and hopes it will Continew   both the Childering are in good health   your Papa is very well   there has been nothing partickular wanting   the man took the theodiled and paid 7£ without the Least scruple   he did not need a Chain and never spoke of instruments   I supos it was a Contrivance of Osburns   Mr. Grea has imployed Heatly to Make his Machine but told him he had first offered it to you – but that you were ingaged with Dr. Roebuck – he said he knew that was trew and that you were both dammed mean and impertinet for you were making an ingine of which he was the Contriver and had given a hint of it to the Docter who he said was as great a scoundrel as your self for doing such a thing   he would make your Pattant of no avail for the Moment he knew it was going he would put a stop to it – he said a great deal more impertinet stuf about the Docters bad useage of him till Mr. Grea said he knew your Carikter to be good and could not believe him for if you would have maid his ingin he had never imployed another   he had always harde the Doctor had an unexceptionable Carikter and beged he would talk of the busniss he was sent about for – I wrote this in case you and the Dr. thought of any means of silencing him – for he may find a great many that will believe him   writ if you d[on't] come home this week   Dr. Hill was [   ] informed   Yours

MARGRET WATT'

13. Extracts from letters from John Roebuck junior to James Watt

Dr. Roebuck's son also wrote to give news of Hateley's engine. Hateley evidently did not understand the basic principles of Watt's improvements, but made the complaint (later to be reiterated by others) about the vagueness of Watt's specification. Many years later, John Roebuck junior wrote to confirm that his father had never claimed any share in Watt's invention, but had only helped to develop it; he also suggested that Hateley had tried to pirate Watt's engine while in Roebuck's employ.

(i) JOHN ROEBUCK JUNIOR TO JAMES WATT,
19 October 1769, Doldowlod.

'. . . Hately boasted much to me of his Engine. he said that it went 15 or 18 in the Minute, told Mr. Grieve that it went much better for cold Water about it. That if you worked by rarefied Air it was his, also if you put any water about it it was his contrivance. But he does not understand the Principles [tear in paper] of no Alternate Heat or Cold and says that [there are] many things in the specification which you are pleased to call Inventions that nobody can understand and that if he has made free with your Caracter and my Fathers, that you had more than balanced it by abusing him. . . .'

(ii) JOHN ROEBUCK JUNIOR TO JAMES WATT,
22 November 1796, Doldowlod.

'. . . I perfectly recollect soon after my arrival at Kinneil from France seeing you at Kinneil and that my father told me that he wished I would understand the nature of Fire Engines by paying attention to the Engines on his Colliery and that one was new and lately erected, but that you had invented a Fire Engine of a different Construction from these or any other, which you would explain to me, after I understood the principles and manner of working the Engines at Boness for that you was going to bring Engines of your Invention to be put up at Kinneil and that you was to take out a Patent for your Invention in which he was to have a Share because he was to assist you in finding money necessary in carrying on the business of making Fire Engines according to your invention. I know very well that my Father did render you this assistance about the time of your taking out the Patent – but I never heard my Father in the smallest degree claim any merit or pretend to have any share in inventing, or even in improving upon, your Engine – On the contrary he always represented the whole Invention to be Yours. Joseph Hateley came into my Fathers Service a considerable time after you had been making experiments at Kinneil with your Engine and I recollect that there was one of your small Engines I mean a small Cylinder of Copper and its apparatus or a great Part of [it] lying in one of the Garret Rooms of Kinneil House and I recollect Joseph Hateley with my younger Brothers going frequently to the Top of the House of Kinneil thro the appartments where your Engine lay, under pretence of the prospect which is extensive from the leads on the top of Kinneil House. . . .'

14. 'Comparison of Mr. Blakey's and Mr. Watt's Steam Engines.' *Printed. No date. Doldowlod.*

Another improved steam engine was that of William Blakey, who took out a patent in 1766 for a modified version of Savery's engine, in which, among other things, he used oil on the surface of the water in the steam vessel or cylinder so as to reduce the condensation of steam. But this engine, as the following document demonstrates, was quite different in principle from Watt's engine, and Watt specifically excluded all Savery-type engines from his specification (see above, p. 61).

I. In Mr. *Blakey's* Engine the Cylinder or Steam Vessel is at every Stroke heated by the Steam, and cooled by an Injection of cold Water, and also by the Water that is to be raised which is received into it.

In Mr. *Watt's* Engine the Cylinder is kept always as hot as the Steam whilst the Engine is working, and no cold Water is suffered to enter it.

II. Mr. *Blakey* condenses the Steam in the Cylinder itself by an Injection of cold Water, as usual in the common Fire Engine.

In Mr. *Watt's* Engine the Steam is not condensed in the Cylinder itself, but in another Vessel called *The Condenser*, which only communicates with the Cylinder during the Time of Condensation.

III. Mr. *Blakey* admits Air into his Cylinder on purpose to prevent the Steam from coming into Contact with the cold Water which is in it.

Mr. *Watt* takes the greatest Care to prevent Air from getting into that End of the Cylinder where the Vacuum is intended to be made, and employs Air Pumps to extract any that may accidentally get in.

IV. Mr. *Blakey's* Engine has no Piston, but he employs the elastic force of Steam to press upon the Air which is in his Cylinder, and thereby forces the Water contained below it to pass into another Vessel, from which it rises through a Pipe to the Height required; and it is to be observed, that the Water to be raised enters both the Steam Vessel itself and Two others having open Communications with it.

Mr. *Watt* employs either Atmosphere or the elastic Force of Steam to press upon a Piston which is made Steam Tight, and moves up and down in a Cylinder or other regular formed Vessel; he positively asserts (as may be seen in his Bill, Page 3) that he does not mean to preclude any Engine where the Water to be raised enters the Steam Vessel itself, or any Vessel having an open Communication with it.

V. Mr. *Blakey* condenses his Steam into Water at every Stroke.

Mr. *Watt* proposes in certain Cases to move Engines by the alternate Expansion and Contraction of Steam, in which Cases it is not to be condensed into Water.

Lastly, Mr. *Blakey's* Engine has no Piston, Pump, or Great Lever, but acts in the same Manner as *Savary's* Engine, by the Force of the Steam exerted upon the Surface of the Water to be raised: It differs from *Savary's* by the Interposition of Air and Oil between the Steam and the Water, and in having an Air Vessel in the ascending Pipe.

Mr. *Watt's* Engine resembles the common Fire Engine, and raises Water by a Piston moving in a Cylinder, which Piston is connected with a Lever that works the Pump: It differs from the common Engine and from all others in having the Cylinder always as hot as the Steam, in having no Injection of cold Water into the Cylinder, and in having the Steam condensed in another Vessel, which only communicates with the Cylinder during the Time of Condensation.

Mr. *Watt* has only pointed out the essential Differences between Mr. *Blakey's* Engine and his own, which indeed agree in no Point except in being Steam Engines; but he has purposely avoided making any Comparison of the Two Engines in point of Utility.

15. Extracts from letters from Dr. John Roebuck to James Watt, *1772–75, Doldowlod.*

The following letters from Roebuck to Watt refer to the continued experiments at Boulton's Soho works to develop both the 'Vertical' (reciprocating) and 'Circular' (rotative) engines; reference is also made to Roebuck's assistance in Parliamentary lobbying on the Engine Bill in 1775.

(i) DR. JOHN ROEBUCK TO JAMES WATT,
20 August 1772, Doldowlod.

'. . . Mr. Bolton and Dr. Small have got the Vertical Machine ready for trying as soon as they receive the Quicksilver from London. If the Circular Engine should not answer I should not be discouraged to try the reciprocal one. I think it may be executed with sufficient accuracy to answer. . . .'

(ii) DR. JOHN ROEBUCK TO JAMES WATT,
5 September 1772, Doldowlod.

'. . . By this time Dr. Small and Mr. Bolton will probably be trying the Circular Engine. But as they can not well succeed without a double Atmosphere I told them I considered the Experiment as imperfect. Though if it does not answer their expectation they may condense the Steam in this Circular Machine according to Your original Plan. . . .'

(iii) DR. JOHN ROEBUCK TO JAMES WATT,
12 November 1774, Doldowlod.

'. . . I hope you will have it in your power to inform me that your Willow Piston continues tight after it has been worked some weeks. I do not wonder that your Condenser should be imperfect. I am rather surprized that it should work at all considering the slightness of the materials and its long exposure to the injuries of the Weather – When you have tryed the fusible Mettal in your Circular Machine I shall wish to be informed of the Experiment – You have now effectually established the justness of the Principle on which your Machine is constructed and the generous and spirited Gentlemen [you] are connected with will never suffer it to fail for want of exertion to carry it into execution. . . .'

(iv) DR. JOHN ROEBUCK TO JAMES WATT,
4 March 1775, Doldowlod.

'. . . I have spoken to Mr. William Adam and desired him to give you his assistance in procuring a Bill to secure to you the right of Your Invention of the Fire Engine. And as he has promised me to be so obliging as to enter into a minute detail of it so as [to] be thoroughly master of the Subject which he will defend for you in the House if necessary. I have given you a Card which you will please to seal and deliver to him as early as you can on Tuesday Morning. . . .'

16. Copy of the Minutes of the Parliamentary Committee on Watt's Engine Bill, 1775.
MS. Doldowlod.

A great deal of information about the development of Watt's engine, its principles, performance, and superiority to 'common' engines is provided in the evidence given before the Parliamentary committee on the Engine Bill in 1775.

Martis 11. die Aprilis 1775
Committee on recommittment of Mr. Watt's Engine Bill.
Mr. Adam in the chair.
Present

Mr. Dundas	Mr. Curzon
Mr. Campbell	Mr. Rashleigh
Mr. Grenville	Col. Vaughan
Sir Cecil Wray	

Bill read 1st
2nd ppp

Mr. Joseph Harrison (a Smith at the Soho Manufactory at Birmingham).
Have you ever seen an Engine at Work erected by Mr. Watt at Mr. Boulton's House in Soho?
Yes.
What was the size of the Cylinder?
18 Inches Diameter.
What was the Heighth of the Column of Water?
24 feet.

What was the Diameter of the pump?

   18 Inches and a half.

What was the length of the Stroke in the Cylinder?

   5 feet.

What was the length of the Stroke in the pump?

   6 feet.

What quantity of Coals were consumed in an hour by working the Engine?

   Half a hundred.

What were the Strokes of the Engine in an hour?

   800 – I have counted them – I was at the working of them – I have seen Common Fire Engines work.

Whether the piston descended as quick in this, as in Common Fire Engines?

   Yes – full as quick – The Method took in counting the Strokes was, by fixing up a Machine that had a Communication with the Engine Beam, that moved one Division every stroke of the Engine.

What Quantity of Fuel was used to work a Common Fire Engine of the same Dimensions as this Engine?

   The best Engine we found at Wednesbury consumed 1100 Weight of Coal an Hour – the Diameter of the Cylinder was 42 Inches – the Diameter of the pump was 12 Inches – and the Column of Water that it raised was 204 feet – it made 10 Strokes in a Minute – I took Notice of another Engine at Wednesbury at Mr. Wood's forge, the Cylinder was 6 feet Diameter – it raised a Column of Water 24 feet high and 3 feet Diameter – the Diameter of the Pump was 3 feet – it made about 10 strokes in a Minute – The Men told me they consumed 100 Ton of Coals a Week – I saw the fire, and in my own Opinion I think they must consume that Quantity – They had 2 Boilers [each] 20 feet Diameter.

What was the reason of your examining those Engines?

   Mr. Boulton went along with me, to see whether our Engine was better or worse than theirs – I found that Mr. Watt's Engine did 5 times as much Work with the same Quantity of Coals.

   I have seen one of Mr. Watt's Wheel Engines at Work – The Wheel was 6 feet Diameter – worked by Steam – it moved with great Velocity.

What was the Engine employed on?

   It was erected for Experiments to prove what could be done – Myself and another Man while it was at rest attempted to hold it – but as soon as the Steam was let into it, it was out of our power to hold it – The Wheel weighed upwards of a Ton.

Do you apprehend the Quantity of fire used would be in proportion to Mr. Watt's Beam Engine?

   Yes.

                                     Withdrew.

Matthew Boulton Esq.

   I have one of Mr. Watt's Steam Wheel Engines, and also one of his reciprocating Engines – I have examined many of the Common fire Engines.

In what respect does Mr. Watt's Engine differ from the best Common Engine you have examined?

The best was that at Wednesbury mentioned by the last Witness – The piston of the Common Engine is pressed down by the Weight of the Atmosphere – but that of Mr. Watt's is not – The Condensation of the Steam in the Common Engine is effected by a Jet of cold Water within the Cylinder, but in Mr. Watt's Engine there is no Injection whatsoever of cold Water in the Cylinder.

In what respects is Mr. Watt's Engine better than the Common fire Engine?

The best Common Fire Engine that I have examined has required from 3 to 4 Times the Coal that Mr. Watts does to do the same Work in the same Time – I mean the same Quantity of Water and the same Heighth – 1/4th of Coals in one Engine will produce the same Quantity of Water and to the same Heighth.

How do you account for this different Effect – from the Construction of those different Engines?

In the Common Engine the Cylinder is robbed of a great Quantity of its Heat at every stroke of the piston by the following Causes – 1st. by the great Quantity of cold Water that is injected into the Cylinder to Condense the Steam 2dly. by a small Column of Water that lies upon the Top of the piston in order to keep it Air tight and 3dly. by the Cylinder itself being exposed to the Common Atmosphere.

Is this Engine differently formed from any other Engine?

Yes.

Has it the Advantage of saving 3 or 4 Times the Quantity of Coals?

Yes. This Engine produces a greater Effect with 1/3d. of the Steam than the Common fire Engine does with the whole of the Steam – for these reasons 1st. because the Steam is more perfectly condensed in a Vessel separate from the Cylinder in which a much greater Degree of Cold is applied than is convenient or proper to apply in the Common Engine. 2dly. because the Vacuum is not injured by the re-generation of steam from the Hot Water that remains in the Cylinder of the Common Fire Engine, there being no Water whatever admitted into the Cylinder of Mr. Watt's Engine – 3dly. because the piston is pressed down by Steam itself, being a greater power than the Common Atmosphere.

Is the Expence a great deal more than the Common Engine?

It would be rather more Expence – but it would do 4 Times the Work.

How much more Water will be thrown the same heighth with Mr. Watt's Engine, than the old one – the Dimensions being the same?

As 12 to 7.

How many pound to the Square Inch does the Common Engine Work?

I have seen Common Engines work with 4 pound upon the Inch – and others with 8 and 9 pound upon the Inch – but my Opinion is that the Common Engine will not work to Advantage with more than 7 pound upon the Inch – Mr. Watt's Engine works very advantageously with 12 pound upon the Inch.

Suppose 1000£ laid out upon an Engine in the Common Way and 1000£ laid out upon one of Mr. Watt's Engines – Mr. Watt's Engine will do as much Work with 1/3d. or 1/4th. of the Coals.

Withdrew.

Col. Roy.

I have seen many Common Fire Engines and I have examined one more particularly than the rest – The Imperfections of the common Fire Engine is that the Cylinder is alternately hot and cold – that is, the condensation is made in the Cylinder itself – That naturally cools the Cylinder; and in order to re-heat it a greater Quantity of Steam must be produced; in order to produce which a much greater Quantity of Fuel must be used – The piston in the Common fire Engine moves from the pressure of the Atmosphere in the Cylinder, which I understand to have a less Effect than the Effect of Steam.

If the Vacuum was perfect and no part of the Steam was needlessly condensed, how much more Work would the Common Engine do with the same Quantity of fuel?

Making the Comparison by the York-building Engine we found by Examination that the Effects upon a superficial Inch of the piston amounted to 5 pounds and 3 quarters; combining that Effect with the other Circumstances the Quantity of Water necessary to be produced into Steam, and consequently the Quantity of fuel to make that Steam, it appeared that the York-building Engine approached only 1/4th. towards perfection – Supposing the Engine had been compleat I apprehend it would have done 3/4ths. more – supposing it possible to make an Engine compleat.

Are you acquainted with Mr. Watt's Engine?

I have considered the Description Drawings and Model.

In what does it differ from the Common Engines?

It essentially differs in these points, the Cylinder is kept constantly hot, never receiving any cold Water into it, or any other cold Body – The Condensation being made in a Vessel apart – and which has no Communication with the Cylinder but in the Moment of Condensation – It is Steam that presses down the piston.

What advantage may be gained by such Construction?

A great Saving of Steam and Fuel – I never heard of any other Engine so constructed.

Do you think the grant of an exclusive privilege to Mr. Watt for making these Engines would preclude others from making or using the Common Fire Engines?

No – unless they chose to preclude themselves by adopting the more, instead of the less perfect Engine.

Whether those 7 principles are new Inventions of Mr. Watt – or whether they have been used by others?

Applied in a new way by Mr. Watt.

Withdrew.

Adjourned till Tomorrow.

Mercurii 12 die Aprilis 1775.
Committee on Mr. Watt's Engine Bill.
Mr. Adam in the Chair.

Present

Col. Campbell   Mr. John Goodrick
Mr. Brett   Mr. Tuffnell
Mr. Greville   Col. Vaughan
Col. Livingstone   General Conway

Mr. Robert Mylne.

I have considered Mr. Watt's Invention upon Steam Engines, in consequence of his Communication of the Discovery to me; and I think, that in the present State of the Engines, as to their perfection, it is one of the most meritorious Discoveries, of the present Time.

In what respect, is it superior to those now in use?

All fire Engines, hitherto made known, practised, and used, are reducible into 2 Classes; those, which having a vibrating Balance Beam and pump Gear, operate by the pressure of the Atmosphere upon a piston. Of this Class, is Mr. Newcomens Invention; and almost all the Engines, made use of throughout these Countries. The other Class, are those, which have Vessels emptied, so as to suck Water up to one Heighth; and, by the expansive power of Steam, pressing upon *that* Water, so as to force *it* up to a further Heighth. Of this Species or Class, are the Inventions of the Marquis of Worcester, Captain Savery, and Mr. Blakey. – Mr. Watt's Invention is totally different from both of them; for, being possessed of a vibrating Beam, a piston, and a Cylinder, it does not operate by the Pressure of the Atmosphere. The Merit of his Invention consists of this, that the Cylinder, in which the Steam is received, is not obliged to undergo, an immediate and instantaneous Alteration, from Heat to Cold, by the Condensation of Steam within it, as in the Common Engines; but, being maintained in a constant degree of Heat, equal to that of Steam, in its most perfect State, the Condensation is performed, in a Vessel separate, appropriated for that purpose only. By this Means, the want of perfection, in the Cylinders of Common Engines, is avoided; which may be described in this manner. In the working of an Engine, were it to be quite perfect, it ought to be perfectly hot at one Instant of Time – equal to Steam; at another period, it ought to be perfectly cold, in order to make a compleat Vacuum. This alternate Operation, being to be produced, from 16 to 20 Times in a Minute, (considering how *they* are formed, and with what surrounded,) is in the Nature of Things impossible. From *this*, arises the Imperfections of the Common Fire Engine – and on the contrary, the perfection of Mr. Watt's Invention. Because the Cylinder according to Mr. Watt, has a Body of Steam round it, to preserve the Metal perfectly hot; it has no Water over the piston, which damps the power of the Steam, in the Inside of the Common Engine; there is no Injection of cold Water into it, which, in the Common Engine is raised to a certain Degree of Heat, and together with the Water produced from the Steam condensed, produces Steam, so as to render the

Operation, in part ineffectual, at the very period of Time, it ought to operate. Another Merit, that Mr. Watt's Invention has, by employing the pressure of the power of Steam upon the piston, instead of the Atmosphere, is an Excellence, superior to the Common Engine, in the Degree of 12 to 7.

Are you acquainted with the Effects produced by Mr. Watt's Machine?

In no other respect, than by a deduction of reasoning, from what is already done; having the Care of Fire Engines under me.

What additional Advantage, would any person have, who erects and uses Mr. Watt's Engine?

From having an Engine of the Common sort, working every day throughout the Year, and finding by Experience the difficulty of producing Steam, at the Instant of Condensation, in the most perfect State of Elasticity, and how much Coal is expended for that purpose, I conclude that the Savings of fuel must be very great; and of Course, the Expence of Working the Engine, considerably reduced. From a Survey of its Construction, and the Estimate of the Expence of erecting such Engines, I judge, there will be no saving in the first Instance, of erecting one of them – but that the whole consists in the current Annual Expence, as near as I can judge, of 300£ per Annum in 400£ and so proportionably, for any other Sum.

What is the Expence of a Common Engine whose Steam Vessel is 18 Inches Diameter?

In London, such an Engine, without the Building, and including all the Pump Work, will cost about 1400£. This is not an Opinion, or an Estimate, but from Experience, within these 3 or 4 Years.

What is the Diameter of the pump, and the Heighth of the Lift of such an Engine?

The Column of Water is 18 Inches and a half Diameter, and 32 feet perpendicular Heighth; and the Merit of the Engine, indifferent.

What is the Expence of an Engine upon Mr. Watt's principle, whose Steam Vessel is 18 Inches Diameter, in London?

If a reference is had to the Diameter of the Cylinder – and taking *that* as a Standard between the 2 Engines, I apprehend such an Engine, will cost double that of another; but at the same Time will do double the Quantity of Work.

Is it possible to give a greater force by Steam, than what is done by this Machine of Mr. Watt's?

I have no Conception of it.

What National Encouragements had the Marquis of Worcester, Captain Savery, and Doctor Newcomen for their respective Machines?

I have an Act of Parliament by me, granting to the Marquis of Worcester, an exclusive right to his Invention of a Steam Engine for 99 Years, which I believe expired about 1762. Captain Savory in a publication in 1701 says, that he then had a patent for it, and also an Act of Parliament; but for what Term of years, I am not able to say.

Withdrew.

Mr Boulton [produces Copy of a Clause from the Act of Parliament granting a patent to Mr. Savory for his Invention]

I have examined the parliament Rolls relative to Mr. Savory's Engine – and have obtained a Copy of the Act from which I took the above Clause.

What is the Term of Years granted by the Act?

Captain Savory's patent was dated in July 1698 – on the 20th. March following the Act is dated giving him a further Term of 21 Years, which Amounts in all to 35 Years.

When was Dr. Newcomen's Improvement by the Lever made?

I believe about the same Time as Mr. Savory's.

Had Dr. Newcomen any parliamentary Encouragement.

I have heard he had a patent.

Withdrew.

Mr. Samuel Garbett – called,

I have seen Mr. Watt's Wheel-Engine.

Can you judge of the powers of that Engine?

I cannot – I went to see it merely as a Matter of Curiosity – It moved with a very considerable Velocity.

Could a Man have stop't it?

Two Men attempted to prevent its going into Motion, and could not – I saw a little Vessel connected with this Wheel by a pipe, in which Steam was produced – The Steam that went from this Vessel to the Wheel in my Opinion moved it – It might be connected with any other Machine.

If connected with any other Machine should those 2 Machines be considered as one combined Machine?

If they were moved by that Steam they must be considered as one.

By what means do you apprehend the Steam operated upon this Engine?

I know nothing of it.

Withdrew.

Mr. Mylne – called in again.

Here is a Machine, particularly and Mechanically described by the Bill, under Consideration of this Committee, to which it is proposed to give, under that Description, a Monopoly. The powers of this Machine, are proposed to give Motion to a second Machine, whose first Movement is a Wheel. The question I ask, is, whether these 2 Machines thus combined, are in the description of Mathematicians or Mechanicks, one combined Machine?

I apprehend that the circular Movement produced from Steam, such as is described in the Bill, is in itself not a Machine but a power raised – a Machine, consists not only of the power raised, but of the Business done. When this power raised, is applied to any purposes of Life, it becomes a distinct Machine; – and the Merit of this Invention consists in the Application of Steam, to create a circular Motion.

May a Pump be worked by a Wheel?

No Doubt.

Whether the Application of this Invention, to work a pump with this wheel, so invented, would be considered as a combined Invention, to be included to be within this Act.

No Doubt of it.

Whether the Application of this Invention, to work any Machine by this Wheel so invented, would not also be included within the Monopoly granted by this Act?

No Doubt of it: – and when it is granted, I don't think it a sufficient reward, for such a Meritorious Invention.

What is the fluid Body raised by the force of Steam on one side?

Quicksilver: – or a Body similar in its Nature. Quicksilver is the only Body, I know of at present, that has been used in the Engine.

Must not the Steam be of the density of 2 Atmospheres to raise the Quicksilver 14 Inches?

I can't answer that Question.

Withdrew.

17. 1775 Steam Engine Act

ANNO REGNI DECIMO QUINTO GEORGII III. REGIS
CAP. LXI.

AN ACT FOR VESTING IN JAMES WATT, ENGINEER, HIS EXECUTORS, ADMINIS-TRATORS, AND ASSIGNS, THE SOLE USE AND PROPERTY OF CERTAIN STEAM ENGINES, COMMONLY CALLED FIRE ENGINES, OF HIS INVENTION, DESCRIBED IN THE SAID ACT, THROUGHOUT HIS MAJESTY'S DOMINIONS, FOR A LIMITED TIME.

WHEREAS His Most Excellent Majesty King George the Third, by His Letters Patent, under the Great Seal of Great Britain, bearing date the fifth day of January, in the ninth year of his reign, did give and grant unto JAMES WATT, of the city of Glasgow, Merchant, his executors, administrators, and assigns, the sole benefit and advantage of making and vending certain engines, by him invented, for lessening the consumption of steam and fuel in fire engines within that part of his Majesty's Kingdom of Great Britain called England, the Dominion of Wales, and the Town of Berwick upon Tweed, and also in his Majesty's Colonies and Plantations abroad, for the term of fourteen years; with a Proviso obliging the said JAMES WATT, by writing under his hand and seal, to cause a particular description of the nature of the said invention to be inrolled in His Majesty's High Court of Chancery, within four months after the date of the said recited Letters Patent:

AND WHEREAS the said JAMES WATT did, in pursuance of the said Proviso, cause a particular description of the said engine to be inrolled in the said High Court of Chancery, upon the twenty-ninth day of April, in the year of our Lord one thousand seven hundred and sixty-nine, which description is in the words and form, or to the effect following; that is to say,

MY METHOD of lessening the consumption of steam, and consequently fuel, in fire engines, consists of the following principles:

FIRST, That vessel in which the powers of steam are to be employed to work the engine, which is called the Cylinder in common fire engines, and which I call the Steam Vessel, must, during the whole time the engine is at work, be kept as hot as the steam that enters it; first, by enclosing it in a case of wood, or any other materials that transmit heat slowly; secondly, by surrounding it with steam, or other heated bodies; and, thirdly, by suffering neither water, nor any other substance colder than the steam, to enter or touch it during that time.

SECONDLY, In engines that are to be worked wholly or partially by condensation of steam, the steam is to be condensed in vessels distinct from the steam vessels or cylinders, although occasionally communicating with them; these vessels I call Condensers; and, whilst the engines are working, these condensers ought at least to be kept as cold as the air in the neighbourhood of the engines, by application of water, or other cold bodies.

THIRDLY, Whatever air or other elastick vapour is not condensed by the cold of the condenser, and may impede the working of the engine, is to be drawn out of the steam vessels or condensers by means of pumps, wrought by the engines themselves, or otherwise.

FOURTHLY, I intend in many cases to employ the expansive force of steam to press on the pistons, or whatever may be used instead of them, in the same manner as the pressure of the atmosphere is now employed in common fire engines; in cases where cold water cannot be had in plenty, the engines may be wrought by this force of steam only, by discharging the steam into the open air after it has done its office; (which fourth article the said JAMES WATT declares, in a note affixed to the specification of the said engine, should not be understood to extend to any engine where the water to be raised enters the steam vessel itself, or any vessel having an open communication with it).

FIFTHLY, Where motions round an axis are required, I make the steam vessels in form of hollow rings, or circular channels, with proper inlets and outlets for the steam, mounted on horizontal axles, like the wheels of a water-mill; within them are placed a number of valves, that suffer any body to go round the channel in one direction only; in these steam vessels are placed weights, so fitted to them as entirely to fill up a part or portion of their channels, yet rendered capable of moving freely in them, by the means hereinafter mentioned or specified: When the steam is admitted in these engines, between these weights and the valves, it acts equally on both, so as to raise the weight to one side of the wheel, and by the re-action on the valves, successively, to give a circular motion to the wheel, the valves opening in the direction in which the weights are pressed, but not in the contrary; as the steam vessel moves round, it is supplied with steam from the boiler, and that which has performed its office may either be discharged by means of condensers, or into the open air.

SIXTHLY, I intend, in some cases, to apply a degree of cold, not capable of reducing the steam to water, but of contracting it considerably, so that the engines shall be worked by the alternate expansion and contraction of the steam.

LASTLY, Instead of using water to render the piston or other parts of the engines air and steamtight, I employ oils, wax, resinous bodies, fat of animals, quicksilver, and other metals, in their fluid state.

AND WHEREAS, the said JAMES WATT hath employed many years, and a considerable part of his fortune, in making experiments upon steam, and steam engines, commonly called fire engines, with a view to improve those very useful machines, by which several very considerable advantages over the common steam engines are acquired: but upon account of the many difficulties which always arise in the execution of such large and complex machines, and of the long time requisite to make the necessary trials, he could not complete his intention before the end of the year one thousand seven hundred and seventy-four, when he finished some large engines as specimens of his construction, which have succeeded so as to demonstrate the utility of the said invention:

AND WHEREAS, in order to manufacture these engines with the necessary accuracy, and so that they may be sold at moderate prices, a considerable sum of money must be previously expended in erecting mills, and other apparatus; and as several years, and repeated proofs, will be required before any considerable part of the publick can be fully convinced of the utility of the invention, and of their interest to adopt the same, the whole term granted by the said Letters Patent may probably elapse before the said JAMES WATT can receive an advantage adequate to his labour and invention:

AND WHEREAS, by furnishing mechanical powers at much less expense, and in more convenient forms, than has hitherto been done, his engines may be of great utility in facilitating the operations in many great works and manufactures of this kingdom; yet it will not be in the power of the said JAMES WATT to carry his invention into that complete execution which he wishes, and so as to render the same of the highest utility to the publick of which it is capable, unless the term granted by the said Letters Patent be prolonged, and his property in the said invention secured, not only within that part of Great Britain called England, the Dominion of Wales, the Town of Berwick upon Tweed, and his Majesty's Colonies and Plantations abroad, but also within that part of Great Britain called Scotland, for such time as may enable him to obtain an adequate recompence for his labour, time, and expence:

TO THE END, therefore, that the said JAMES WATT may be enabled and encouraged to prosecute and complete his said invention, so that the public may reap all the advantages to be derived therefrom in their fullest extent, may it please Your Most Excellent Majesty (at the humble petition and request of the said JAMES WATT) that it may be enacted,

AND BE IT ENACTED, by the King's Most Excellent Majesty, by and with the advice and consent of the Lords Spiritual and Temporal, and Commons, in this present Parliament assembled, and by the authority of the same, that, from and after the passing of this Act, the sole privilege and advantage of making, constructing, and selling the said engines, herein-before particularly described, within the Kingdom of Great Britain, and his Majesty's Colonies and Plantations abroad, shall be, and are hereby declared to be, vested in the said JAMES WATT, his executors, administrators, and assigns, for and during the term of twenty-five years; and that the said JAMES WATT, his executors, administrators, and assigns, and every of them, by himself and themselves, or by his and their deputy or deputies, servants or agents, or such others as he the said JAMES WATT, his executors, administrators, and assigns, shall at any time agree with, and for no others, from time to time, and at all times, during the term of years herein-before mentioned, shall and lawfully may make, use, exercise, and vend the said engines, within the Kingdom of Great Britain, and in his Majesty's Colonies and Plantations abroad, in such manner as to him the said JAMES WATT, his executors, administrators, and assigns, shall in their discretions seem meet; and that the said JAMES WATT, his executors, administrators, and assigns, shall and lawfully may have and enjoy the whole profit, benefit, commodity, and advantage, from time to time coming, growing, accruing, and arising, by reason of these his said inventions, for the said term of twenty-five years, to have, hold, receive and enjoy the same, for and during and to the full end and term of twenty-five years as aforesaid; and that no other person or persons within the Kingdom of Great Britain, or any of his Majesty's Colonies or Plantations abroad, shall, at any time, during the said term of twenty-five years, either directly or indirectly, do, make, use, or put in practice, the said inventions, or any part of the same, so attained unto by the said JAMES WATT as aforesaid, nor in anywise counterfeit, imitate, or resemble the same; nor shall make, or cause to be made, any addition thereunto, or subtraction from the same, whereby to pretend himself or themselves the inventor or inventors, devisor or devisors thereof, without the licence, consent, or agreement of the said JAMES WATT, his executors, administrators, or assigns, in writing under his or their hand and seal, or hands and seals, first had and obtained in that behalf, upon such pains and penalties as can or may be justly inflicted on such offenders for their contempt of this Act; and further, to be answerable to the said JAMES WATT, his executors, administrators, and assigns, according to law, for his and their damages thereby occasioned.

PROVIDED ALWAYS, and be it hereby declared, that nothing in this Act contained shall extend, or be construed to extend, to prejudice or hinder any person or persons from making or using any fire or steam engine, or any particular contrivance relating to the same, which is not at present of the invention of the said JAMES WATT, or which has been publickly used or exercised by any other person or persons before the time of the date of the said Letters Patent herein recited; but that all such engines and contrivances which are not at present of the said invention of the said JAMES WATT, or are not particularly specified in this Act, shall be and remain to the publick, and to the respective inventors thereof, as if this Act had never been made, anything herein contained to the contrary notwithstanding.

PROVIDED ALSO, that every objection in law competent against the said patent shall be competent against this Act to all intents and purposes, except so far as relates to the term hereby granted.

PROVIDED ALWAYS, that if the said JAMES WATT, his executors, administrators, or assigns, or any person or persons who shall at any time, during the said term of twenty-five years, have or claim any right, title, or interest, in law or equity, of, in, or to the power, privilege, or authority, of the sole use and benefit of the said invention, shall make any transfer or assignment, or pretended transfer or assignment, of the said liberty or privilege hereby granted, or any share or shares of the benefit or profits thereof; or shall declare any trust thereof to or for any number of persons exceeding the number of five, or shall open, or cause to be opened, any book or books for public subscriptions, to be made by any number of persons exceeding the number of five, in order to the raising any sum or sums of money, under pretence of carrying on the said liberty or privilege hereby granted; or shall by him or themselves, or his or their agents or servants, receive any sum or sums of money whatsoever, of any number of persons exceeding in the whole the number of five, for such or the like intents or purposes; or shall presume to act as a corporate body; or shall divide the benefit of the liberty or privileges hereby granted, into any number of shares exceeding the number of five; or shall commit or do, or procure to be committed or done, any act, matter, or thing whatsoever, during such time as such person or persons shall have any right or title, either in law or equity, which shall be contrary to the true intent and meaning of an Act of Parliament, made in the sixth year of the reign of His late Majesty King George the First, intituled, 'An Act for the better securing certain Powers and Privileges intended to be 'granted by His Majesty, by Two Charters, for Assurance of Ships and Merchandises at 'Sea, and for lending Money upon Bottomry, and for restraining several extravagant and 'unwarrantable Practices therein mentioned;' or in case the said power, privilege, or authority, shall at any time become vested in or in trust for more than the number of five persons, or their representatives, at any one time, otherwise than by devise or succession, (reckoning executors and administrators as and for the single persons whom they represent, as to such interest as they are or shall be entitled to in right of such their testator or intestare); that then, and [in] every of the said cases, all liberties and advantages whatsoever hereby granted shall utterly cease, determine, and become void; anything herein-before contained to the contrary thereof in anywise notwithstanding.

AND BE IT FURTHER ENACTED by the authority aforesaid, that this Act shall be adjudged, deemed, and taken to be a public Act; and shall be judicially taken notice of as such by all judges, justices, and other persons whomsoever, without specially pleading the same.

18. James Watt to his Father, 8 May 1775, Doldowlod.

The following letter from Watt to his father gives some impression of the Parliamentary struggle and suspense during the procedure on the Engine Bill, and indicates the business prospects opened up by its passage.

'Dear Father

'After a series of various and violent Oppositions I have at last got an Act of Parliament vesting the property of my new Fire engines in me and my Assigns, through out Great Britain and the plantations for twenty five years to come, which I hope will be very beneficial to me, as there is already considerable demand for them.

'This affair has been attended with great Expence and anxiety, and without many friends of great interest I should never have been able to carry it through, as many of the most powerfull people in the house of Commons opposed it. It has been in parliament ever since the 22nd of February, which is a very long time to be kept in suspence.

'I shall be obliged to stay here a few days longer after which I return to Birmingham to set about making some Engines that are ordered, after which I intend to give myself the happiness of seeing you and the dear Children. I would have wrote to you oftener but had so little that was pleasing to communicate that I contented myself with letting you know of my welfare through the Channel of our friends –

'My warmest wishes and Affection ever attend you, and God render your Age comfortable is the prayer of your ever, affectionate and

Dutiful Son

JAMES WATT

remember to Jean, her Brother and sisters and all other friends – '

19. Matthew Boulton to James Watt, *March 1776, Doldowlod.*

Watt's hopes of rising orders for his engine are echoed in this letter from Boulton. John Wilkinson, the famous ironmaster, whose newly-patented boring engine produced cylinders of the requisite accuracy, was also one of the first customers for their engine. The Cornish tin and copper mines, remote from coalfields, were a particularly good prospect for their fuel-saving engines.

'Dear Sir,

'I rejoyce at the well doing of Willey Engine [a blowing engine for John Wilkinson's blast-furnaces at New Willey, near Broseley, Staffs.] as I now hope and flatter my self that we are at the Eve of a fortune. I wish to see you at Soho as soon as possable. there are many things want you and I find my self exceedingly hurryed. People are daily coming to see the Engines. Cornwall begins to Enquire how we go on. I will reserve particulars untill I see you but if you dont come in a very few days I shall be gone to London being obliged to attend parliament again. . . .'

20. J. Smeaton to Boulton and Watt, 5 February 1778, Birmingham Reference Library.

Watt always expressed respect for the engineering abilities of John Smeaton, who, though most famous for his civil engineering achievements, had, in fact, like Watt, been originally an instrument-maker and displayed a similar interest and ability in experimental philosophy and mechanics. After having made great advances in water-wheel design and construction, as a result of carefully controlled experiments, he then turned to the steam engine. Realizing, like Watt, the great loss of heat and fuel through condensation in the cylinder of the Newcomen engine, he carried out prolonged experiments during the early 1770s which enabled him to double its efficiency. In the following letter, however, he paid generous tribute to Watt's superior genius in producing the idea of a separate condenser. At the same time, he showed considerable acumen in his opinion that, whilst Watt's engine would undoubtedly be successful in areas where coal was dear, it would be slow in displacing common engines on the coalfields themselves.

'. . . I now come to that part of your Letter wherein you so generously and genteely offer, in consideration of the good will you bear me, the whole profitts accrewing from an application of a condenser, by a separate Vessell, in eny form I chuse, to one Engine: provided it be an Engine employed upon a Colliery: and that I do not make a free gift of these profitts to my Employer . . . you have only to say what share of the saving, my employer shall pay you; for as to myself, as I expect he will pay me for my Labour and Time in the ordinary course of business; this is all the advantage I desire, or indeed will receive: the gratification of my own Curiosity, and the chance of doing you a piece of service, being to me a very sufficient inducement to take the Extra pains attending the Application of Your Idea, of condensing in a separate vessel from the cylinder: the whole profit of which is justly your due and which alone I look upon as a greater Stroke of invention, than has appeared, since Newcommen reduced the Engine to its present form: And had Mr. Watt been more early in producing this Idea to Light, he had spared me a vast deal of pains, and the expending a capital upon the Subject. – You are pleased to express your wishes, that we might have had a long conversation which might have been for our mutual Interest; I can only say you do me wrong if you suppose I have ever acted in opposition to your inventions: I have on the other hand always spoke of them as I have thought of them, that is Exceedingly ingenious, and adapted to remove the very thing that was the real rub, or stop to the improvement of Newcommens Engine beyond a certain Degree; and that is the condensation of the Steam in getting in to the Cylinder; this I not only fully investigated by my Experiments, upon my own Experimental Engine, that was set agoing in 1770; but had strongly suspected it to be very considerable from the performance of the New River Engine; that was designed in 1767, built in 68, and sett to work the beginning of the year 1769: but could never have believed this condensation to have been in the degree it is, had I not afterwards fully investigated the same by means of my Experimental Engine, which was built on purpose to get into the full light of those facts, that in the New River Engine turned out so Extraordinary: And it would seem indeed as if I had designed the N River Engine on purpose, to bring out something that was not then known. I considered the Stoppage of the Collumn of water every Stroke, as well as the putting the great Lever piston, heavy chains and Rodds twice into motion from Rest every stroke, as a great loss of momentum: I therefore Determined not only to work the Engine Slower, with larger pumps, and all the load it would bear; but to reduce the velocity of the Collumn still more, by taking a Stroke of 9 feet at the Piston End, while the pumps Stroke was only 6 feet. This naturally gave me a very long and Narrow Cylinder; from whence I expected this benefit would arise, that every part of the Steam contained, being nearer the Surface of the Cylinder (which long before that I had found to be the principal cause of condensation) would be more readily condensed; in consequence a lesser quantity of Injection water would serve, the Cylinder itself be more heated, and the vacuum more perfect. Under all these appearances of advantage, I ventured to give the piston a neat burthen of water, amounting to 10¼ lb pr Inch. Having once seen a common engine struggle under this burthen, I thought myself quite secure under those advantages; But how great was my surprize and Mortification to find, that instead of needing less Injection water than common, that the Injection pump, (which was calculated to afford as much injection water as common, in proportion to the Area of the Cylinder, with a sufficient overplus to answer all imaginable wants) was unable to supply the Engine with [enough] Injection; and that a couple of men were obliged to be clapped on to raise the injection water quicker by hand, in order to keep the Engine in motion; At the same time that the Cylinder was so cold that I could keep my hand upon any part of it, and bear my hand for a length of time in the Hot well: by good luck however, the Engine performed the Work it was expected to do, as to the raising of Water; but the Coals by no means answered calculation. The Injection pump being Enlarged, it was in a State of doing Business. Many smaller Experiments I tryed, but without any good Effect, till I altered the fulcrum of the Beam; and reduced the load upon the piston from lb 10¼ to 8¼ lb pr Inch; under this Load, tho it shortned Stroke at the pump End, yet it went so much more quick as not only to raise more Water, but consumed less coals took less injection water pr stroke, the Cylinder became Hott, and the injection water came out at 180° of Farenheit, and the engine in every respect not only did its work better, but went more pleasantly. This at once convinced me that a considerable degree of condensation took place in entering the Cylinder, than I had gained by the increase of Load: in short this single Alteration seemed to have unfettered the Engine: but in what degree this condensation took place under different circumstances of heat; and where to strike the medium, so as upon the whole to do best, was still unknown to me. But resolving, if possible, to make myself master of the Subject, I immediately resolved to build a small Engine at home, that I could Easily convert in to different shapes for Experiments; and which Engine was very near ready to sett to work, when I first heard of Mr. Watts Idea of condensing in a separate vessel from the Cylinder; which was from Mr. Mitchel; to whom I immediately replyed, Mr. Watt has certainly found out where the shoe pinches: it is a thought that will either do very well, or not at all; but for my own part I dont see how it can be applyed; that is how to get rid of the Air generated in the Condenser; for, if he snifts *that*, he will loose as much Steam in that vessel, as he will save in the cylinder. And it was not till the Summer following viz of 1771 (in which Interim I had made some fundamental Experiments) that I learnt from Mr. Watt himself; that his method of clearing the condenser of Air and Water, was by an Air pump: but as he then acquainted me, that having built an Engine at Kinneal, they had for the present been brought to a Stop, by a difficulty they had found and had not satisfactorily removed; viz that of keeping the piston Steam tight, without Water, and Mr. Watt being then wholly occupied in a different concern, and the Engine laid aside for the present, to be resumed or not as it might happen, I determined to prosecute my original intentions of finding out the true *Rationale*, and making the best of Newcommens Construction; nor did I hear anything more of a revival of Mr. Watts Scheme, till your Application to Parliament the spring of the year 1775: in which interim I had not only produced the Longbenton Engine, which was the first in consequence of those Improvements but two capital Engines: one for Cronstadt and the other for Chacewater, the Great parts whereof were then prepared, and ready for putting together.

'This history of what I have done in the Engine way, cotemporary with Mr. Watt, tho it is drawn out much beyond the length that I expected or intended, when I dipt my pen in Ink, for this part of the subject of my letter; yet I have been thus minute in order to shew you clearly, that what I have done has not been in any degree in the way of competition, or of supplanting, or of lessening the value of the improvements that you have made; On the contrary, I may rather say that having brought my scheme to bear, and being in full possession of a series of valuable improvements, by which that very important machine could be made to act with a double effect to what is done in common; and with a considerable advantage over Every one on the common principle that I have yet seen; before you had brought yours into the light of a practicable bearing and being thus in the way of professional Business; you come and *oust me*, by further improvements, rather than I *you:* Yet so far have I been from being envious or discontented hereat, that the very first opportunity that offered, after you have brought your affair to a practicable light, I mean after your Exhibition before Parliament, I recommended it . . . Nothing since has [been] offered to me, but Collery Engines (and you see you have already driven me to the Coal pitt hill) till the present one: and in case you dont think it proper to give me leave to use your Condenser I am determined immediately to putt my friend into your Hands.

'Thus my dear Sirs, you will see I have ever conducted myself towards you, not as an Ennemy but as a disinterested friend. True it is, I have often said; and *still* must say, that though I have no doubt, but that your Scheme will answer a very valuable End where coals are Dear; And be of little consequence upon the coal pitt Hill: Yet I dont as Yet know the quantume of validity of your Scheme; how to calculate, where to draw the line, or how to recommend. If I knew this, It might be in my power, more directly to promote your Interest. The fact is, that to this hour, I have no account upon which I can depend, of the actual performance upon a fair and well attested Experiment of anyone of your Engines. . . .'

21. Extracts from letters from James Watt to Matthew Boulton, *April 1781, Doldowlod.*

Watt's ideas for improving the steam engine originally included, as we have seen, a scheme for a 'steam-wheel', to give 'circular' or rotative motion. But this ultimately proved unsuccessful and all the early Boulton and Watt engines were reciprocating engines. By the late 1770s, however, Watt's mind was turning towards the conversion of reciprocating into rotative motion. He intended to make use of the ancient device of the crank, but his ideas were revealed by a workman and patented by Wasbrough and Pickard in 1779 and 1780. Watt's disgust and fury is revealed in the following letters, which also refer to his device for avoiding use of the crank – his famous 'sun-and-planet' motion, which he patented in 1781.

(i) JAMES WATT TO MATTHEW BOULTON,
2 April 1781, Doldowlod.

'. . . I think before we expose ourselves to any lawsuit in the affair of the Crank we ought to have the advice of able counsel – I dare not make my new scheme lest we be betrayed again. I believe we had best take the patent first – '

(ii) JAMES WATT TO MATTHEW BOULTON,
4 April 1781, Doldowlod.

'I wrote to you on Sunday with the copy of Specification of the Crank engine. . . . I saw Mr. Dearman yesterday and today he says that his Brother Sampson wishes to have the affair made up with us and wants an engine from us for some place or other – [The Dearmans were local ironfounders, evidently associated with Pickard and Wasbrough in erecting their rotative engine on Snow Hill, Birmingham.] I told him that we could clearly prove priority of Invention and also that our man had communicated the invention to their man – I mentioned to him that you were now taking the advice of Counsel that both you and I were loth to begin a lawsuit but that we would not lose the use of our own invention. I told him that so far as concerned me I should not give them any trouble if they would gratis communicate the benefits of the patent, which he said Sampson seemed inclined to – I told him however that I could not answer for your sentiments – They seem to want to make some terms of our permitting them to use our Engine which I object to if general. It ought to be special for each Engine and not compulsive on us – They ought to give us a Charte Blanche for we can use the other invention which is as good as theirs and besides have a very good Claim on theirs which I fancy would lay it open and they have no claim on our Engine.'

(iii) JAMES WATT TO MATTHEW BOULTON,
9 April 1781, Doldowlod.

'If we agree with the Snow Hill Gentlemen they will propose having the use of our engines on certain terms and though we should have the unlimited and free use of the crank it will not be equivalent as they will constantly avail themselves of our improvements, which the machine much needs, and will interfere with us in every shape where these rotatives are concerned now if we make no bargain with them but pursue the spiral they cannot make small Engines to any advantage and by the superiority of our Engines we shall cut them out in most cases.'

(iv) JAMES WATT TO MATTHEW BOULTON,
13 April 1781, Doldowlod.

'. . . Quere would it be right to consult the Attorney general or Sollicitor on the subject of their patent – and if by an application to the privy council in a matter of so much Injustice would not the patent be taken away.

'I think without they make an unconditional Transfer of part of their patent right we should come into no terms with them but try to lay the patent open, all england will be better customers to us than they can be and moreover what is every bodies business is nobodys and they being broke up we might have the business principally to ourselves –

'I have contrived another sort and have sett Joseph to make it. I believe it will do very well for going one way – but does not admit of reversing. . . .'

(v) JAMES WATT TO MATTHEW BOULTON,
21 April 1781, Doldowlod.

'I have wrote the inclosed so that you may show it to whom you will. It contains my real sentiments but expressed more forcibly than I chuse to do to you. This attempt of Mathew's has awakened every angry passion in my breast and I cannot bring myself to submitt to such an indignity let what will be the consequence – I cannot conceive that the Commissioners will attempt to exert the arm of power over us, why should we be sacrificed to M[atthew] W[asborough]. You should exert any interest you have to prevent any thing of that kind, and I realy think we should take advice whether the matter cannot be brought before the privy council who have a right to take away the patent if proved to be injuriously obtained – They have no power over our Act, thank God! I have given orders to make preparations for trying the spirals but nobody is entrusted yet but Joseph – I believe the best policy will be to publish it to all the world and depend only on the profits for our engines, by which means we shall knock up this damned Company altogether and gett quit of their nonsense. they deserve such a punishment from us – In the mean time perhaps some of the servants of the Office want grease or have got it of the wrong sort. . . .'

(vi) JAMES WATT TO MATTHEW BOULTON,
22 April 1781, Doldowlod.

'. . . I think still that working to our own Contrivance while claimed by Mr. Wasborough

7 Wax medallion of Matthew Boulton by Peter Rouw. *By courtesy of the Assay Office, Birmingham*

8 Soho Factory, Birmingham, 1798. *By courtesy of the Birmingham Reference Library*

9 Sectional model of Watt's single-acting engine. *By courtesy of Birmingham University*

will be an indelible disgrace and will hurt us in any future Contest we may have for them – Therefore we should either adopt our own Engine with the spiral or resign the business – but if you find yourself so circumstanced as you say you are that you dare not refuse then let them pay Mr. Wasborough and have done with him and lett the Engine be erected under our direction or Mr. Smeatons. with the latter I will go hand in hand nay I will do more I will submit to him in all mechanical matters but I will by no means submit to go on with thieves and puppies whose knowledge and integrity I contemn. Though I am not so saucy as many of my Countrymen I have enough of innate pride to prevent me from doing a mean action because a servile prudence may dictate it – If a king should think Matt Wasborough a better engineer than me I should scorn to undeceive him, I should leave that to Mathew. The conviction would be the stronger as the evidence would be undeniable – so much for heroics!

'Now for common sense. I agree with you that no terms are to be made with these people except full and unbounded liberty to make as many Crank engines as we please without any other Considerations than cloaking their theft which then and not before it will be our interest to do – I will never meanly sue to a thief to give me my own again without I have nothing left behind – As it now stands I have enough left to make their patent tremble and shall leave no mechanical stone unturned to agrieve them. I will do more. I will publish my inventions by which means they will be entirely precluded because they must be fools indeed that will pay *them* for what they can have for nothing. I am very ill with a head ache therefore can write no more than passion dictates – My very best regards to Mr. Smeaton.

'Adieu, most sincerely your's JAMES WATT

take care to sign no paper that acknowledges Wasboroughs right – nor none in conjunction with him.'

(vii) JAMES WATT TO MATTHEW BOULTON,
28 April 1781, Doldowlod.

'. . . being thoroughly convinced in my own mind that they invented no part of it. they got the hint of the Crank from Ned Ruston who had it from a lathe and the weight they got from us via Cartwright –'

(viii) JAMES WATT TO MATTHEW BOULTON,
April 1781, Doldowlod.

'. . . When you asked my opinion about the proposals made to you by the victualling Office I gave my consent not from the desire of gain but because you seemed to wish it and that I thought it wrong to refuse our assistance to national service, and therefore resolved patiently to submit to my share of the labour and to work through it as speedily as bad health would permit me; but I never consented *nor ever will* to hold the candle to Mr. Wasborough – Had I esteemed him a man of Ingenuity and the real inventor of the thing in question I should not have made any objection, but when I know the contrivance is my own and has been stole from me by the most infamous means and to add to the provocation a patent surreptitiously obtained for it, I think it would be descending below the Character of a *Man* to be any ways aiding or assisting to him or to his pretended inventions – I am perswaded that the Commissioners of the victualling office will not desire us to join with persons who from what I have said must be to the last degree disagreable to us from which circumstance alone I should expect the machine to be badly managed and every bad Consequence to ensue – I wish to live in quiet and will not expose myself to the vexations which must unavoidably arise in such a case – If the Honourable Board judge it proper to employ us for the whole Machinery, i.e. the Engine and the Rotative motion I will exert every faculty which God has given me to make it a good Machine and I am sure that I have the ability to do so without interfering with Wasboroughs patent, granting We should not have justice done us in a court of law whither I foresee that case must come – I say I know from experiment that the other contrivance which you saw me try performs at least as well and has in fact many advantages over the crank – I think therefore that you should propose to the honourable Board to undertake the direction of the whole and provided you can agree with them about the customary premium for the savings by our Engine You should do the whirligig part for Love – If this proposal should not be accepted, I beg of you to decline having any concern with it, and leave the field Clear to Mr. Wasborough. we may perhaps gain more by so doing than we can lose as I assure you I have a very mean opinion of the mechanical abilities of our Opponents they have committed many gross errors in such of their works as I have occasion to know about and we may get honour by rectifying their mistake. . . .'

22. Matthew Boulton to James Watt,
   21 June 1781, Letter Book M, 1781–83, Assay Office Library, Birmingham.

Boulton's optimism in regard to the industrial possibilities of steam power was now greatly stimulated, since rotative engines could be used not merely in mining operations but for driving machinery of all kinds. Manufacturers were 'Steam Mill Mad', and Boulton therefore urged Watt to patent as soon as possible.

'I think I understand all the elements of the horizontal, Eliptical, Ecliptical, Conical, comical Rotative Mill, and a good Mill it is therefore I beg neither it nor any of the others which we have made (and nobody else) may be lost for want of Secureing, as the people in London, Manchester, and Birmingham, are Steam Mill Mad, and therefore let us be wise and take the advantage. I dont mean to hurry you into any determination, but I think in the course of a Month or two we should determine to take out a patent for certain methods of producing Rotative Motions from the vibrating or reciprocating Motion of the Fire Engine remembering that we have 4 Months to discribe the particulars of the invention.'

23. 1781 Specification of Patent

This patent was granted on 25 October 1781 and the specification was enrolled on 23 February 1782. The specification describes five different methods whereby the reciprocating action of the steam engine might be converted into rotative motion without using a crank, thus evading Pickard's patent. Of these, the fifth – the 'sun-and-planet' gear – was the one principally adopted by Boulton and Watt, until expiry of Pickard's patent, when they turned over to the crank.

SPECIFICATION OF PATENT, OCTOBER 25TH, 1781, FOR CERTAIN NEW METHODS OF APPLYING THE VIBRATING OR RECIPROCATING MOTION OF STEAM OR FIRE ENGINES, TO PRODUCE A CONTINUED ROTATIVE OR CIRCULAR MOTION ROUND AN AXIS OR CENTRE, AND THEREBY TO GIVE MOTION TO THE WHEELS OF MILLS OR OTHER MACHINES.

TO ALL TO WHOM these presents shall come, I, JAMES WATT, of Birmingham, in the county of Warwick, Engineer, send greeting.

WHEREAS His Most Excellent Majesty King George the Third, by His Letters Patent under the Great Seal of Great Britain, bearing date at Westminster, the twenty-fifth day of October, in the twenty-second year of his reign, did give and grant unto me, the said JAMES WATT, my executors, administrators, and assigns, His especial licence, full power, sole privilege and authority, that I, the said JAMES WATT, my executors, administrators, and assigns, should, and lawfully might, during the term of years therein expressed, make, use, exercise, and vend, within that part of his Majesty's Kingdom of Great Britain called England, his Dominion of Wales, and Town of Berwick upon Tweed, my invention of 'CERTAIN NEW METHODS OF APPLYING THE VIBRATING OR RECIPROCATING 'MOTION OF STEAM OR FIRE ENGINES, TO PRODUCE A CONTINUED ROTATIVE OR 'CIRCULAR MOTION ROUND AN AXIS OR CENTRE, AND THEREBY TO GIVE MOTION 'TO THE WHEELS OF MILLS OR OTHER MACHINES;' in which said recited Letters Patent is contained a Proviso obliging me, the said JAMES WATT, by an instrument in writing, under my hand and seal, to cause a particular description of the nature of my said invention, and in what manner the same is to be performed, to be inrolled in His Majesty's High Court of Chancery within four calendar months next and immediately after the date of the said Letters Patent, as in and by the said Letters Patent, relation being thereunto had, may more at large appear:

NOW KNOW YE, That in compliance with the said Proviso, I the said JAMES WATT do hereby declare that the nature of my said invention, and the manner in which the same is to be performed, is particularly described and ascertained in manner and form following (that is to say): the Fire or Steam Engines whose vibrating or reciprocating motions are to be converted into rotative motions by any or all of the five methods hereinafter described, may be constructed either upon the principles of the steam engines called Newcomen's Fire or Steam Engines, (which have been hitherto most commonly used), or, more advantageously, upon the principles of those newly improved steam or fire engines of my invention, (the sole use and property of which was granted to me by his present Majesty's Royal Letters Patent, dated in the ninth year, and by an Act of Parliament made and passed in the fifteenth year of his reign): or the said engines may be constructed in any other manner or mode wherein a piston or any other part of the said steam or fire engine has a vibrating, alternating, or reciprocating motion: therefore, as for the aforesaid purpose no peculiar construction is required in those parts of the steam or fire engines which concur in and are necessary for the producing the power or active force of the engine, and its vibratory or reciprocating motion, and as steam or fire engines are common and well known machines, it is not necessary to enter into any description of them; I proceed to explain my said newly invented methods of applying the vibrating or reciprocating motions of steam or fire engines to produce a continued rotative or circular motion round an axis or centre, and thereby to give motion to the wheels of mills and other machines, which methods are five in number, and are described as followeth:

IN THE FIRST OF THESE METHODS I employ the power of the steam engine, either directly, or by the intervention of a lever or levers, to pull, push, or press a friction wheel or pulley against the lateral surface of a wheel fixed obliquely upon the primary axis, shaft, or wheel which is to receive the rotative motion; which lateral surface of the said oblique or inclined wheel is represented by the section of a hollow cylinder ABC in the drawing, No. 1, figure 1st, cut or sawn off at the angle of sixty-five degrees to its axis, or at any other angle which may be convenient or useful; and the said friction wheel or pulley J is impelled or pulled by the power of the steam engine against the said lateral surface of the inclined wheel AC, in a direction which is in one way parallel to the said primary axis or shaft DE of the said obliquely cut cylinder or inclined wheel AC; therefore the friction wheel J, commencing its motion at the lowest or nearest part C of the said inclined wheel AC, continues to move in the aforesaid direction nearly parallel in one way to the primary shaft or axis DE, and thereby obliges the inclined wheel AC and the primary axis DE to turn round or revolve on their centre until the highest or most distant part A of the said oblique surface of the inclined wheel AC comes into contact with the said friction wheel J, at which point or time the working beam PP, or other moving part of the steam engine, has moved the length of its stroke, and is disposed to return by the common or other machinery used for that purpose, and the inclined wheel or obliquely cut cylinder ABC has made one-half of a revolution on its axis, and the rotative motion of the said inclined wheel ABC is continued in the same direction through the other half revolution by means of the descent of the heavy arch G, which was raised by the power of the steam engine at the same time with the friction wheel J, and which, during the returning motion of the working beam of the steam engine, acts upon the inclined wheel AC on the opposite side of the primary axis DE by means of a second friction wheel H, which is carried by a double lever or carriage GF, whose centre of motion is at K, and to the one end of which the heavy arch G, or any other weight, is attached or suspended, and the velocity which the matter of the wheel or cylinder ABC

90

has acquired serves to continue its rotative motion past the points A and C, where neither the steam engine nor the weight G have much action upon it; and when the point C has again come into contact with or has passed the friction wheel J, the steam engine again commences its action, and the motion is continued, as has been recited, and the mill-work or other machinery which is required to be wrought by this machine is put in motion by the said primary axis, or by the oblique or inclined wheel, or by means of wheels connected with them in the usual manner. This method of producing a rotative motion by means of a friction wheel acting against the lateral surface of a wheel inclined to its axis, admits of many varieties in its mode of application. For I fix the primary axis or shaft either perpendicular or horizontal, or at any other angle of inclination to the horizon which may be required; and I use one or more friction wheels, and I increase or diminish the angle of inclination of the oblique wheel to the primary axis, as the case may require: as therefore I cannot herein represent all those varieties, I have hereunto annexed a drawing or delineation and description of one of the best, (which is applicable to the moving of corn and other similar mills), which drawing is delineated in its true proportions, according to a scale of one-fourth of an inch for each foot of the real machine, being one forty-eighth part of the real size. But it must be remembered that I make the machines larger or lesser, and vary the proportions of their parts, as their uses may require. To shew the easiest method of connecting the said new machinery with the piston of the steam or fire engine, on whatever principle it may otherwise be constructed, I have delineated in red the piston and cylinder of a Newcomen's steam engine, (as being the most commonly used). And as the same mode of connection serves equally for all the four following methods herein described, I have not repeated the drawing of the said steam engine, but have only delineated the parts which in these methods connect the new machinery with the old.

MY SECOND METHOD of producing the aforesaid rotative motion consists in applying the power of the steam engine to pull, push, or press a friction wheel or wheels against the external or internal circumference of a circular, oval or double spiral wheel, fixed upon an axis or shaft in such manner that the said axis or shaft shall not pass through the centre of the said circular, oval or double spiral wheel, but shall be fixed nearer to one side of the circumference than to the other, which therefore I denominate an excentric wheel; and the said action of the steam engine and of the said friction wheel or wheels upon or against the circumference of the said excentric wheel, causes it to make one-half of a revolution, and its motion is continued through the other half revolution by the descent of a weight fixed to or acting upon the said excentric wheel or its shaft, or acting upon another excentric wheel, (fixed to the same shaft), by means of a friction wheel or wheels. This second method also admits of several varieties in its application, of which I have hereunto annexed a delineation of two of the best, shewing the action of the steam engine on the external and also on the internal circumference of excentric wheels, which drawings are delineated and set forth according to their true proportions, by a scale of one-fourth of an inch for each foot of their real size; but the said machines are also made larger or lesser, and the proportions of their parts varied, according to the uses for which they are required. The excentric wheel, whose external circumference is acted upon by the steam engine by means of friction wheels, is moved as follows – (see the drawing No. 2, figure 1st): the steam engine pulls up the frame HDL with the friction wheels F, G, against the external circumference of the excentric wheel AB, which causes it to revolve on its axis towards D, until the point A of the excentric wheel comes to be in the middle between the points of contact of two friction wheels F, G, with the excentric wheel, and the point B has attained the summit of its motion; then the steam engine ceases to act, and the velocity acquired by the excentric wheel AB carries its point B beyond the summit, and the gravity of the unbalanced part of the excentric wheel, which is made equal to half the power of the steam engine, or greater or lesser as may be necessary, causes the excentric wheel to perform the other half revolution, by which motion it pulls down the frame HDL and the end J of the steam engine's working beam, and the point B having past its lowest place, the engine begins to act as before. The action of the steam engine on the internal circumference of an excentric wheel is described as follows: when the engine pulls up the frame DE, (see the annexed drawing, No. 2, figure 3rd), with the friction wheel C, the latter is pressed against the internal circumference of the excentric wheel at H, by which means the wheel is turned round half a revolution, and the point B becomes the vertex; then the engine ceases to act, and the weight of the wheel descending causes it to continue to revolve in the same direction, and completes the revolution, in like manner as has just been described.

My Third Method of producing the said rotative motion is by means of a rod or rods DB, (see the drawing, No. 3, fig. 1st), one end (D) of which is attached or suspended to the end of the working beam of the steam engine, and the other end B to any point of a wheel AEBF, of a circular or any other form, which wheel is fixed at one end of a shaft or axis C; so that by the revolution of the said wheel and the said axis C, the said latter point of fixture or attachment B shall describe a circle round the centre of the said axis, the diameter of which circle shall be equal to the extent of the stroke of the point of the engine's working beam, to which the end D is attached; and the said wheel AEBF is made so much heavier on one side EBF of the centre than upon the other side A, that the said unbalanced weight EBF shall have an action in its descent equal to one-half of the power of the steam engine which works the machine, or more or less as required; or in place of putting the weight in the wheel ABF itself, it is put upon a lever or other wheel fixed to the said shaft CC in any other part, or is fixed in any other manner which may serve to make the wheel continue its motion during the return of the piston of the steam engine; and this machine is used as follows: when the point or pin B, which connects the rod DB with the wheel, is a little on either side of the lowest part of its revolution, the steam engine pulls the rod DB, and thereby obliges the wheel to make one half of a revolution, and the unbalanced weight EBF of the wheel, or such other weight as acts upon it during the return of the steam engine, makes the wheel complete its revolution, as has been already recited in the other methods. This third method also admits of several varieties in the mode of execution, for the wheel is sometimes placed so as to turn vertically, (as in the drawing), and sometimes to turn horizontally, and also at other inclinations to the horizon, and the balancing weight is also placed in various situations: I have therefore delineated only one of the most simple and perfect of these methods in the drawing No. 3, hereunto annexed, which is laid down by the

92

same scale with the other drawings of the preceding methods, but the size of the machine must be greater or lesser, and the proportions of its parts varied, according to its use.

MY FOURTH METHOD of producing the aforesaid rotative motion consists in employing two steam engines to produce a rotative motion in one and the same axis or shaft by any of the aforesaid three preceding methods, or by that which is hereinafter described, and in applying these two steam engines in such manner that the second engine shall begin to act when the first engine has made the said shaft revolve upon its axis one third part of a revolution or thereabouts, and consequently, by the action of both the engines, the shaft makes two third parts of a revolution, and its motion is continued through the remaining one third part of the revolution by the action of a weight properly placed, by which means the rotative motion is maintained in a more equal manner than can be done by a single steam engine. I also apply this method to move two separate or distinct shafts or axles, which are connected in their action by wheel-work, or otherwise, so that they both must revolve the same number of turns in the same time. As this fourth method must admit of many varieties in its application to any or all of the three preceding or the following methods, all which may be easily understood by explaining its application to one of them, I have only delineated in the drawing No. 4 its application to the third method, as being the most simple, and I have laid down the said drawing according to the same scale with the others. The motion of this machine is explained as under. The pin G of the connecting rod BG, (see drawing No. 4, fig. 2nd), having passed its lowest point, the working beam B ascends by the power of the steam engine to which it belongs, and, by its action on the pin G through the rod BG, causes both the wheels and their common axis to revolve, until the end of the rod BG arrives at the point C, at which time the end of the rod AC, (which is attached to the further wheel), is arrived at the point K, (that is, a point of the further wheel directly behind K in this view), and the centre of gravity of the weight, or heavy sides of the wheels G, J, C, has passed its vertex or highest part at F, and begins to descend towards K; the rod BG continues to act upon the wheels until it arrives at F, where it ceases to act, and the motion is continued only by the gravity of the heavy side of the wheels, or by any other properly disposed weight, until the point C of the further wheel has past its lowest place, and comes into the position directly behind G, when the steam engine belonging to the rod AC begins to act, and continues the motion until it arrives again at C, when the revolution is completed, and the rod B acts as before: the heavy sides of the wheels, or any other weight used to continue the motion, ought to be equal to one-half of the power of one of the engines, but may be greater or lesser, as suits.

MY FIFTH METHOD of producing the aforesaid rotative motion, (delineated in the drawing No. 5 hereunto annexed), is performed by means of a toothed wheel E, (fig. 1st), fixed upon the end of the shaft or axis F which is to receive the rotative motion; which wheel E is acted upon and made to revolve by means of a second toothed wheel D of an equal, or greater, or lesser diameter, which is firmly fixed to or connected with a rod AB, (the other end of which is attached or hung to the working beam BC of the steam engine, or is otherwise connected with the piston of the said engine), in such manner that the said wheel D cannot turn round on its own axis or centre; and by means of a pin A, which is fixed to, or in the centre of the wheel D, and enters into a groove or circular channel in the large wheel GG, (or by any other proper means), the wheel D is confined so that it cannot recede from the wheel E, but can revolve or turn round the wheel E without turning on its own axis or centre, and the motion is performed as follows. The wheel D being nearly in the position of the pricked circle HH, and so that its centre shall be a little towards either side of the perpendicular line passing through the centre F; the steam engine, by means of the connecting rod BA, pulls the wheel D upwards, and, as its teeth are locked in the teeth of the wheel E, and it cannot turn on its own axis, it cannot rise upwards without causing the wheel E to turn round upon its axis F: when the wheel D is raised so high that its lower edge is come into contact with the upper edge of the wheel E, the engine has completed its stroke upwards, the piston of the engine is disposed to return, and the wheel E, continuing to turn round in virtue of the motion it had acquired, it carries the wheel D past the vertex or highest part, and the gravity of the wheel D, or of the connecting rod AB, or of any other weight connected with them, causes the wheel D to descend on the other side of the wheel E, and thereby continues the motion it had impressed upon it, whereby the wheel D completes its revolution round E; and when the two wheels D and E have equal numbers of teeth, the wheel E makes two revolutions on its axis for each stroke of the engine; and in order that the said motion may be more regular, I fix to or upon the shaft or axis FML, (fig. 2nd), or to or upon some other wheel or shaft to which it gives motion, a heavy wheel or flyer, to receive and continue the motion communicated to it by the primary movement. AND BE IT REMEMBERED, that in all cases where heavy wheels or swift motions are not otherwise necessary to the uses to which any of the four preceding methods herein described may be applied, a flyer or heavy rotative wheel should be applied to them to equalise their motion. In figures 3rd and 4th, I have delineated the application of this method to a wheel CC, fixed upon the primary axis, and having teeth upon its inside circumference, which is acted upon by the wheel E in the manner which has just been recited; but as the wheel E has only half the number of teeth that the wheel CC has, the wheel CC will make only one revolution for every two strokes of the engine. The drawings of this fifth method are also delineated according to a scale of one-fourth of an inch for every foot of the real size of the machine. Figures 1st and 2nd are adapted to a stroke of six feet long, and figures 3rd and 4th to a stroke of three feet long, but I make the machines larger or lesser, and also make such variations in their structure as may serve to accommodate them to their use; as I alter the proportional diameters of the two wheels, and I place the primary axis either horizontally, perpendicularly, or inclined, and I make the wheels of an elliptical, oval, or other form; and sometimes, in place of the wheel D, I use a straight row of teeth or pins fixed to the connecting rod AB, which take hold of the teeth of the said wheel E, and cause it to revolve; some point of the connecting rod being guided by a pin moving in a groove, so as to keep the teeth or pins always engaged in the teeth of the wheel E, and also to keep the teeth of the wheels always engaged in one another instead of the wheel GG and its groove. I use a strap of leather or a link of iron, or other proper material, (such as is drawn at JK), which embraces the axis M or F, and the pin A, and connects them together, and keeps them at their proper distance from each other; and I also make the two wheels E and D without any teeth, but with rough surfaces, so that D turns E by the friction of their circumference alone. BE IT REMEMBERED, that though I have described all these motions as derived or produced from the motion of the end of the working beam of the steam engine, they may also be derived from the vibrating motion of any other part of the steam or fire engine which is found convenient: the end of the working beam appears at present to be the best adapted for that purpose: any or all of these methods admit of the machines being moved with rotative motions in either direction, that is, either right-hand ways or left-hand ways about, according as the motion is commenced in either of these directions respectively.

IN WITNESS whereof I have hereunto set my hand and seal, the thirteenth day of February, in the year of our Lord one thousand seven hundred and eighty-two.

JAMES WATT.

Signed, sealed, and delivered, being first duly stamped, by the within named JAMES WATT, in the presence of

N. BENNETT, Clerk to Mr. Davis of Penryn.

BENJ. COLLETT, Servant to Mr. Davis.

INROLLED in His Majesty's High Court of Chancery, the twenty-third day of February, in the year of our Lord 1782, being first duly stamped according to the tenor of the Statutes made in the sixth year of the reign of their late Majesties King William and Queen Mary, and in the seventeenth year of the reign of His Majesty King George the Third.

JOHN MITFORD.

1. P. Y.

24. James Watt to Gilbert Hamilton, 5 March 1782, Doldowlod.

Watt was more cautiously optimistic than Boulton about the prospects of the rotative engine, and feared, indeed, that Boulton's 'foolish vaunting' was helping to raise up piratical competitors, such as the Hornblowers ('Horners').

'. . . I have lately taken out a patent, for producing Rotative or Circular Motions from reciprocating fire Engines, by 5 new methods and we are likely to have many orders in that way for Mills etc. – I am also taking out a patent for the use of our Engines in Ireland, and another for certain Improvements on them in England, the latter is principally intended to close all the holes we can think of by which we may be intruded upon which we have much need to do as in consequence of Mr. B[oulton]s foolish vaunting of our profits the minds of all projectors are bent upon rivalling us and in these Matters conscience is quite out of the Case.

'The noise about the Horners Engine has quite subsided but will probably be renewed if they should be able to make a successfull exhibition, but in the latter case I believe we have a very good suit against them; but the ways of the law are dark and intricate and by no means desirable.'

25. James Watt to Matthew Boulton, *13–16 February 1782, Doldowlod.*

In addition to the patent for rotative motion, Watt also prepared to take out another in 1782, comprising further improvements [26], but was fearful lest he should be again forestalled.

'. . . I hope you will spur on Mr. Handley with the patent and at the same time cause enquire what new patents are now going through the Office for I do not think that we are safe a day to an end in this enterprizing age. Ones thoughts seem to be stolen before one speaks them. It looks as if Nature had taken up an adversion to Monopolies and put the same things in several peoples heads at once to prevent them, and I begin to fear that she has given over inspiring me as it is with the utmost difficulty I can hatch any thing new . . .'

26. 1782 Specification of Patent

This patent was granted on 12 March 1782 and the specification was enrolled on 4 July 1782. It included a number of 'new improvements': (i) the use of the expansive principle (with a valve to cut off steam before completion of the piston stroke, thus economizing on fuel), together with various possible devices for equalizing the expansive power; (ii) a double-acting engine, utilizing steam pressure alternately above and below the piston, thus increasing the power and ensuring a more regular motion in rotative engines; (iii) a double or compound engine, comprising two engines, each with its cylinder and condenser, which could either work independently or be combined in such a way that steam employed in the first cylinder could be used expansively in the second; (iv) a toothed rack-and-sector connecting the piston-rod and beam, to permit 'push' as well as 'pull' in double-acting engines (a device produced by Watt prior to his parallel motion); (v) a steam-wheel or rotative engine of similar principle to that patented in 1769.

SPECIFICATION OF PATENT, MARCH 12TH, 1782, FOR CERTAIN NEW IM-
PROVEMENTS UPON STEAM OR FIRE ENGINES FOR RAISING WATER, AND
OTHER MECHANICAL PURPOSES, AND CERTAIN NEW PIECES OF MECHANISM
APPLICABLE TO THE SAME.

TO ALL TO WHOM these presents shall come, I, JAMES WATT, of Birmingham, in the
county of Warwick, Engineer, send greeting.

WHEREAS His Most Excellent Majesty King George the Third, by His Letters Patent,
under the Great Seal of Great Britain, bearing date at Westminster, the twelfth day of
March, in the twenty-second year of His reign, did give and grant unto me, the said JAMES
WATT, His especial licence, full power, sole privilege and authority, that I, the said JAMES
WATT, my executors, administrators and assigns, should, and lawfully might, during the
term of years therein expressed, make, use, exercise, and vend, within that part of his
Majesty's Kingdom of Great Britain called England, his Dominion of Wales, and Town of
Berwick upon Tweed, my invention of 'CERTAIN NEW IMPROVEMENTS UPON STEAM OR
'FIRE ENGINES FOR RAISING WATER, AND OTHER MECHANICAL PURPOSES, AND
'CERTAIN NEW PIECES OF MECHANISM APPLICABLE TO THE SAME;' in which said
recited Letters Patent is contained a Proviso obliging me, the said JAMES WATT, by an
instrument in writing under my hand and seal, to cause a particular description of the
nature of my said invention, and in what manner the same is to be performed, to be inrolled
in his Majesty's High Court of Chancery within four calendar months after the date of the
said Letters Patent, as in and by the said Letters Patent, relation being thereunto had, may
more at large appear.

NOW KNOW YE, That in compliance with the said Proviso, I the said JAMES WATT do
hereby declare that the nature of my said invention and the manner in which the same is to
be performed, is described and ascertained in manner and form following, (that is to say):
my said new improvements on steam or fire engines, and my said new pieces of mechanism
applicable thereto, are described as followeth. But to prevent misunderstandings and cir-
cumlocutions or tautology, I shall first explain the meaning of certain terms which are used
in this Specification. FIRST, the cylinder or steam vessel is that vessel wherein the powers
of steam or air are employed to work the engine, of whatever form it may be made, though
it is most commonly made cylindrical. SECOND, the piston is a moveable diaphragm,
sliding up and down or to and fro in the cylinder, and fitted to it exactly, on which piston
the powers of steam or air are immediately exerted. THIRD, the condensers are certain
vessels of my invention, in which the steam is condensed, either by immediate mixture with
water sufficiently cold or by contact with other cold bodies, which condensers are situated
either in that part of the cylinder itself which the steam never enters, except to be condensed
or reduced to water, or they communicate with the cylinder by means of pipes, which are
opened or shut at proper times, or those pipes called eduction pipes, which lead to the air
pumps or other contrivance for carrying off the condensed steam, air, and water of injection,
are themselves used for that purpose. FOURTH, the air and hot water pumps are pumps
or other contrivances which serve to extract the air and heated water from the cylinders and

condensers. FIFTH, the working beam is a double-ended lever, a wheel or wheels, or other machinery establishing the means of communicating the power from the piston to the pump work, or other machinery to be wrought by the engine.

MY FIRST NEW IMPROVEMENT in steam or fire engines consists in admitting steam into the cylinders or steam vessels of the engine only during some certain part or portion of the descent or ascent of the piston of the said cylinder, and using the elastic forces, wherewith the said steam expands itself in proceeding to occupy larger spaces, as the acting powers on the piston through the other parts or portions of the length of the stroke of the said piston; and in applying combinations of levers, or other contrivances, to cause the unequal powers wherewith the steam acts upon the piston, to produce uniform effects in working the pumps or other machinery required to be wrought by the said engine: whereby certain large proportions of the steam hitherto found necessary to do the same work are saved. To explain which principle or improvement, I have delineated a section of a hollow cylinder at figure 1st in the annexed drawing. The said cylinder is perfectly shut at the lower end by its bottom CD, and also at the upper end by its cover AB; the solid piston EF is accurately fitted to the said cylinder, so that it may slide easily up and down, yet suffer no steam to pass by it: the said piston is suspended by a rod or rods GH, which is capable of sliding through a hole in the cover AB of the cylinder, and its circumference is made air and steam-tight by a collar of oakum, or other proper materials, well greased, and contained in the box O; and near the top of the cylinder there is an opening J to admit steam from a boiler. The whole cylinder, or as much of it as possible, is inclosed in a case MM containing steam, or otherwise is maintained of the same heat with boiling water, or of the steam from the boiler. These things being thus situated, and the piston placed as near as may be to the top of the cylinder, let the space of the cylinder under the piston be supposed to be exhausted or emptied of air, steam, and other fluids; and let there be a free passage above the piston for the entry of steam from the boiler, and suppose that steam to be of the same density or elastic force as the atmosphere, or able to support a column of mercury of thirty inches high in the barometer. Then I say that the pressure or elastic power of the said steam on every square inch of the area or upper side of the piston, will be nearly fourteen pounds avoirdupois weight, and that if the said power were employed to act upon the piston through the whole length of its stroke, and to work a pump or pumps, either immediately by the piston rod prolonged, or through the medium of a working beam or great lever, as is usual in steam engines, it would raise through the whole length of its stroke a column or columns of water, whose weight should be equal to ten pounds for each square inch of the area of the piston, besides overcoming all the frictions and *vis inertiæ* of the water and the parts of the machine or engine. But supposing the whole distance from the under side of the piston to the bottom of the cylinder to be eight feet, and the passage which admitted the steam from the boiler to be perfectly shut when the piston has descended to the point K two feet, or one-fourth of the length of the stroke or motion of the said piston, I say that when the piston had descended four feet, or one-half of the length of the stroke, the elastic power of the steam would then be equal to seven pounds on each square inch of the area of the piston, or one-half of the original power; and that when the piston had arrived at the point P, the power of the steam would be one-third of the original power, or four pounds and two-thirds of a pound on each square inch of the piston's area; and that when the piston had arrived at the bottom or end of its stroke, that the elastic power of the steam would be one-fourth of its original power, or three pounds and one-half pound on each square inch of the said area. And I further say, that the elastic power of the steam at the other divisions marked in the lengths of the said cylinder, are represented by the lengths of the horizontal lines or ordinates of the curve KL, also marked or delineated in the said cylinder, and are expressed in decimal fractions of the whole original power by the numbers written opposite to the said ordinates or horizontal lines. And I also say, that the sum of all these powers is greater than fifty-seven hundred parts of the original power multiplied by the length of the cylinder; whereby it appears that only one-fourth of the steam necessary to fill the whole cylinder is employed, and that the effect produced is equal to more than one-half of the effect which would have been produced by one whole cylinder full of steam, if it had been admitted to enter freely above the piston during the whole length of its descent; consequently that the said *New* or *Expansive Engine* is capable of easily raising columns of water whose weights are equal to five pounds on every square inch of the area of its piston, and that with one-fourth of the contents of the cylinder of steam. AND BE IT REMEMBERED, though, for example's sake, I have mentioned the admission of one-fourth of the cylinder's fill of steam, (as being the most convenient), that any other proportion of the fill of a cylinder, or any other dimensions of the cylinder, will produce similar effects, and that in practice I actually do vary these proportions as the case requires. And also, in some cases, I admit the required quantity of steam to enter below the piston, and I pull the piston upwards by some external power against the elastic force of the steam from the boiler, which then always freely communicates with the upper part of the cylinder, and which produces similar effects to those described. But the powers which the steam exerts being unequal, and the weight of the water to be raised, or other work to be done by the engines, being supposed to resist equally throughout the whole length of the stroke, it is necessary to render the whole acting powers equal by other means.

I PERFORM THIS, FIRST, by means of two wheels or sectors of circles, one of which is attached to the pump rods, and the other to the piston rod of the engine, and which are connected together by means of rods or chains, or otherwise, so that the levers whereby they act upon one another decrease and increase respectively during the ascent or descent of the piston, in, or nearly in, the ratios required. This method, mechanism, or contrivance, is delineated in fig. 2nd and also its application to one of my new invented steam engines, the sole use and benefit of which was granted to me by an Act of Parliament passed in the fifteenth year of the reign of his present Majesty. The operation of the engine with this new mechanism added to it is described as follows: The piston A being at its highest place, and the part of the cylinder under it exhausted of steam and air, the regulating valve which admits the steam to enter below the piston being shut, and the valve F which allows steam or air to pass to the condenser GK being open, in order to maintain a good vacuum, the top regulating valve D is opened, and permits the steam from the boiler to enter and act upon the piston, which then begins to descend, and to pull round the wheel or sector of a circle to which it is hung: when the point Q of that wheel has moved to R, the piston has descended two feet, and the point V of the wheel to which the pump rods or other machinery wrought by the engine are suspended, being pulled by the rod, 5, connecting it with the other wheel QRSTU, will have moved through the space VW, the regulating valve D is then shut, so that no more steam may be admitted from the boiler during that stroke, but the piston continues to descend by virtue of the expansion of the steam; and when the point Q is come to the points RSTU, the point V is come to the points WXYZ, respectively describing spaces which are nearly proportional to the powers of the steam at the corresponding points of the descent of the piston: when the piston has made its stroke and is come to the bottom of the cylinder, the regulating valve F is shut, and the valve E is opened, by which means the steam passes from the part of the cylinder above the piston to the part below it, by the pipe C, and, so restoring the equilibrium, permits the piston to ascend to its first position. The regulating valve E is then to be shut, and the exhaustion regulating valve F is opened; the steam rushes into the eduction pipe GG, where it meets a jet of cold water, which enters through the injection pipe H, which is opened immediately before the regulating valve F. The contact of this cold water immediately reduces the steam to water, and produces a vacuum under the piston, and thereby enables the elastic force of the steam to act again upon it, as has been described. Or, instead of injecting cold water into the condenser or eduction pipe itself, I bring the steam into contact with thin plates or pipes of metal, which have their external surface cooled by contact with water or other cold matter: the condensed steam, the injection water, and the air which entered with it, or any other air which has entered by other means, proceed by the eduction pipe to the air pump K, and, passing the valves of its bucket or piston, are retained and lifted up by it on the return of the stroke, and thereby are thrown into the hot water pump J, which by the next stroke raises up and delivers them into the atmosphere, from whence part is returned into the boiler to supply its consumption of water, and the remainder is conveyed away for any other purpose, or runs to waste. Another Variety of this method of equalizing the power is delineated in figure 3rd. The piston is suspended to the arch A, by means of a chain, or otherwise, and the pump rod is suspended to the arch B. The primary or cylinder arch A, by means of the arm OP, and the rod or chain OC, acts upon the working beam BC, to the arch of which the pump rods are suspended, by which means, while the piston descends through the equal spaces JK, KL, LM, MN, the pump rod is made to ascend through the unequal spaces DE, EF, FG, GH, which are nearly proportioned to the elastic forces of the steam at the respective points.

My Second Method, or piece of mechanism for equalizing the powers of the steam, is by means of chains, which are wound upon one spiral and wound off another as the piston descends, which spirals are fixed upon two wheels or sectors of circles, to which the chains of the piston and pump rods are attached: and is represented in figure 4th. The piston is suspended from the side A of the wheel AB, and the pump rods from the side C of the wheel DC, which wheel DC being pulled by the chains, J, R, K, the points of its circumference move through the unequal spaces KL, LM, MN, NO (almost exactly proportioned to the powers of the steam), while the points of the circumference of AB move through the equal spaces EF, FG, GH, HI.

MY THIRD METHOD, or piece of mechanism for equalizing the powers of the steam, is by means of a friction wheel or wheels attached to or suspended from one sector or wheel, and acting upon a curved or straight part of another sector, wheel, or working beam. Two modes of this contrivance are delineated in figures 5th and 6th, of which it is only necessary to observe that the pistons of the engines are suspended to the arches AA, and the pump rods to the arches of the working beams BB, and that these contrivances afford the means of equalizing the powers of the steam very exactly.

MY FOURTH METHOD, or piece of mechanism for equalizing the power of the steam, is by causing the centre of suspension of the working beam, or great lever, to change its place during the time of the stroke, whereby the end of the lever to which the piston is suspended becomes longer, and the end to which the pump rods are suspended becomes shorter, as the piston descends in the cylinder. An easy method of doing this is represented in figure 7th: AB represents the working beam; B, the end to which the piston is suspended; A, the end to which the pump rods are suspended; CD, a hollow curve of wood or metal fixed to the lower side of the working beam; E, the end of a friction roller, which rolls between the curve CD and the plane or support FG. This friction roller is divided into three parts, as may be seen in its horizontal view KLM: the two end parts KM, which roll upon the supports FG, are fixed firmly upon an axis; the middle part L, which rolls under the curve CD, can turn round on its axis; therefore, when by the action of the piston on the working beam the end B is pulled downwards, the roller proceeds towards C to the part of the curve which is then the highest, and thereby lengthens the lever by which the piston acts on the pumps, and shortens that by which the pumps resist the cylinder, and that in any ratios which may be required, according to the form of the curve.

MY FIFTH METHOD, piece of mechanism, or contrivance for equalizing the power of the steam, consists in placing upon, suspending from, or fixing to the working beam of the steam engine, or some other beam, wheel, or lever connected with it, a quantity of heavy matter, in such a manner that the said heavy matter shall act against the power of the piston at the commencement of the descent of the said piston, and, as the piston descends, shall gradually move towards that end of the beam to which the piston is suspended, or otherwise shall act in favour of the piston in the latter part of the stroke. Three Methods or Varieties upon this principle are represented at figures 8th, 9th, and 10th. Figure 8th operates by means of a heavy cylinder A, of iron or other material, which rolls in a hollow curve BC on the back of the working beam, and will consequently change its place, and proceed towards the cylinder end of the beam, as the piston descends. In figure 9th, the same is performed by a heavy weight of iron or other matter, A, fixed above the beam or wheel, so that its centre of gravity at beginning the motion lies nearer the pump end of the beam than the centre of suspension of the beam, whereby it acts against the piston, and at last comes to be at the same side of that centre, with the end to which the piston is suspended, and thereby acts in its favour. Figure 10th shows a method of fixing the working beam itself, so as to perform in some degree the office of the weight in figure 9th. Figure 11th shows a Fourth Method, whereby I perform the same thing, by causing a quantity of water or other liquid to oppose the ascent of the piston in the beginning of the stroke, and to assist it in the latter part. A and B represent two cylinders or other formed vessels filled with water, or some other liquid, above their pistons C and D, the rods of which are fixed (or suspended) to the working beam of the engine, or such secondary or auxiliary beam as may be applied to this use, on the opposite sides of its axis. These cylinders are open at bottom and at top, and at the top part they are connected together by a lander or trough, close or open: when the piston descends, it raises the opposite end of the working beam and the piston of cylinder B, and thereby causes the water it contains to run over into A, the piston C of which, becoming loaded thereby in proportion as piston D rises, gradually comes to assist the piston of the steam vessel in the latter part of its motion. These cylinders, containing water, may be either placed below the working beam or above it, or may be suspended to a secondary working beam, constructed for that purpose, or necessary for some other use connected with the piston rod or pump rod, or other part, and placed so that the water cylinders may be out of the engine-house, in some place where it may be found to be more convenient.

MY SIXTH METHOD, piece of mechanism, or contrivance for equalizing the powers of the steam, consists in employing the surplus power of the steam upon the piston in the first parts of its motion, to give a proper rotative or vibratory velocity to a quantity of matter which, retaining that velocity, shall act along with the piston, and assist it in raising the columns of water in the latter part of its motion, when the powers of the steam are defective: two methods by which I perform this are delineated in the drawing of the engine, figure 12th. The heavy fly XX is put in motion by means of a pinion or smaller wheel Y fixed upon its axis, and the teeth of which pinion or smaller wheel are acted upon by the toothed sector QQ, fixed upon the arch of the working beam; or the said fly is by other means connected with the motion of the said working beam. When the piston pulls down the end of the working beam, the toothed sector QQ gives motion to the pinion, and thereby gives velocity to the fly; and when the descending or ascending velocity of the arch or sector of the working beam comes to be less than the velocity which the pinion and fly have acquired, then the velocity of the fly continuing, causes the pinion to act upon the sector in its turn, and assist the powers of the steam, until its velocity is spent, or the piston has reached the bottom of the cylinder; and the said fly operates in like manner during the ascent of the piston, but turns then in the contrary direction. In the Second Variety of this method, a fly, or heavy wheel, is put into a continued rotative motion by a crank, by any of the rotative motions for which I have obtained his Majesty's Royal Letters Patent, dated in the present year of his reign, or by any other means which shall or can produce a continued rotative motion. And the said rotative machinery is connected with or joined to either end of the working beam, to or with the piston rod itself, to or with the pump rods, or to or with any other moving part of the engine or pump rods which is found proper. In the drawing, figure 12th, T, U, W, V, V represent the application of my Fifth Method of producing rotative motions from steam engines, as a method or contrivance for equalizing the power of the steam. The piston being at the top of the cylinder, and the working beam in the situation delineated, the engine begins its stroke, and, by means of the connecting rod TT, pulls upwards the toothed wheel W, which is fixed to the connecting rod in such manner that it cannot turn upon its own axis, and is confined by means of a link or chain reaching from

102

its centre to the axis of the other toothed wheel U, or is otherwise contrived so that it cannot recede from it; therefore, when the action of the engine pulls the wheel W upwards, it revolves round the other wheel U, and causes it (U) to revolve upon its own axis; and the fly or heavy wheel VV being fixed upon the same axis, it is also put into motion, and because of the great power of the steam in the first parts of the stroke, the fly acquires a great velocity, whereby, through the medium of the two wheels and the connecting rod, it acts upon the working beam, and assists the action of the steam in the latter parts of the stroke: when the piston has completed its stroke downwards, the lower edge of the wheel W has passed over the upper edge or highest part of the wheel U, and, the velocity of the fly continuing, the wheel U acts upon the wheel W, and assists the unbalanced weight of the pump rods in raising the piston to the top of the cylinder. BUT BE IT REMEMBERED, that I (the said JAMES WATT) do not mean that anything which I have herein set forth, touching or concerning this second variety of applying my Sixth Method or contrivance for equalizing the powers of the steam, shall be construed or thought to be intended to preclude any other person or persons from using or applying to the moving, turning, or working of mill-work or other machinery, where continued rotative motions are required, any contrivances or inventions for producing rotative motions from steam engines, provided that such rotative motions or machinery be not of my invention, and such engines be not applied principally or solely to the raising of water: the true intent and meaning of the aforesaid last article of this writing, being to specify the means by which I apply continued rotative motions, as some of my methods or contrivances to equalize the expansive powers of the steam, in engines which are used principally or wholly for the raising of water from mines, rivers, ponds, marshes, lakes, and other places.

MY SECOND IMPROVEMENT upon steam or fire engines consists in employing the elastic power of the steam to force the piston upwards, and also to press it downwards alternately by making a vacuum above or below the piston respectively, and at the same time employing the steam to act upon the piston in that end or portion of the cylinder which is not exhausted; so that an engine constructed in this manner can perform twice the quantity of work, or exert double the power in the same time, (with a cylinder of the same size), which has hitherto been done by any steam engine in which the active force of the steam is exerted upon the piston only in one direction, whether upwards or downwards. This improvement, as applied to one of the steam engines of my invention, is delineated in figure 12th. The lower part of the cylinder B being exhausted of air, steam, and other fluids, the regulating valve F being open, and the regulating valves E and N being shut, the regulating valve D is opened, which admits the steam from the boiler to press upon the upper side of the piston, by the action of which steam the piston descends, pulls down the cylinder end of the working beam, and raises the end to which the pump rods are suspended. When the piston is arrived at the bottom of the cylinder, or end of its stroke, the valve F is shut, and the valve E is opened, which admits the steam under the piston, and at the same time the valve D is shut, which prevents the steam from coming from the boiler into that end of the cylinder, and the other valve N in the upper nozzle or regulator box is opened, which permits the steam to rush from above the piston into the eduction pipe GG, where it meets the jet

H   103 of injection water, which condenses it and produces a vacuum in the upper part of the cylinder, which, destroying the equilibrium, permits the steam under the piston to force it upwards. Then the piston rod being fast in the piston, and having the toothed rack OO fixed to its upper end by means of the teeth thereof, which are engaged in the teeth of the toothed sector which is fixed to or forms a part of the arch QQ of the working beam, or by means of double chains or any other practicable method, the piston or its rod raises the cylinder end of the working beam, and also a heavy weight concealed or contained in the arch thereof, or otherwise fixed, attached, or suspended thereto; which weight ought to be equal, or nearly so, to the force or power of the steam when acting upon the piston in the ascending direction. When the piston has arrived at the summit of its motion, the regulating valves E and N are to be shut, and the regulating valves D and F opened, whereby the piston again commences its motion downwards, as has been described; and, during the descent of the piston, the weight QQ, fixed or suspended to the working beam, assists the power of the steam on the piston, in raising the column of water in the pumps, or in working other machinery. In figure 13th is delineated a front view of the cylinder, and a section of the condenser of this engine, wherein the pipes which convey the steam from the boiler, and to the condenser, are properly distinguished and explained. This improvement permits the engine to be used either with the uniform exertion of the whole power of the steam on the piston, both in the descent and ascent; or, by making the weight of the columns of water in the pumps, or the resistance of other machinery which it may be required to work, equal to the full power of the steam upon the piston, when acting in one direction only, and the weight on the working beam equal to half that power, it may be used as a double expansive engine, and wrought in the manner I have herein set forth in the description of my First Improvement. And in such case the Fourth, Fifth, and Sixth contrivances herein described for equalizing the powers of steam, are peculiarly applicable to this mode of constructing the engine. Wherefore I have delineated the two varieties which I have described of the Sixth method as applied to this engine. AND BE IT REMEMBERED, that either or both of them may be used at the same time, though only one is strictly necessary, and that any other two or more of the aforesaid Six contrivances, or of the varieties thereof, may be applied to one engine at the same time; that is to say, such of them whose nature admits of such combination.

MY THIRD IMPROVEMENT on steam or fire engines consists in connecting together, by pipes or other proper channels of communication, the steam vessels and condensers of two or more distinct steam engines; each of which has its separate working beam and other constituent parts of a steam engine, or is otherwise so constructed that it can work pumps or other machinery which are either connected with or are independent of those wrought by the other engine; and which two engines can take their strokes alternately, or both together, as may be required. The construction of the said machine is described as followeth, and an external front view of the steam vessels or cylinders and condenser of the two engines is delineated at figure 14th of the drawings hereunto annexed: the section or side view of the two engines is not delineated, because when viewed in that direction only one of them can be seen, the other being hid by it; and the engine which could be seen would appear the same as the engine delineated in figure 2nd, or, in respect to the working beams, as the engine delineated at figure 12th; – these compound or double engines admitting the application of any of the equalizing machinery which has been hereinbefore described. In figure 14th, the cylinder of the primary engine, No. 1, receives steam from the boiler by the steam pipe 8, 9, which steam enters the cylinder by a regulating valve situated at D, its piston being at the upper end of its stroke; and the part of the cylinder which is below the piston being exhausted, the elastic power of the steam presses down the piston until it arrives at the bottom or termination of its stroke: the regulating valve D is then shut, and the middle regulating valve at E is opened, which admits the steam to enter under the piston, whereby the engine is enabled to raise up the piston to the top of its stroke, where it was at the beginning: the middle regulating valve E is then shut, and the regulating valves F and P are opened: the valve F permits the steam to pass through the eduction pipe N into the perpendicular steam pipe R of the secondary engine, and to press upon its piston, under which piston there is a vacuum. The steam which is or was contained under the piston of the primary engine, No. 1, being of the same density with the atmosphere, or nearly so, will, while the piston of the secondary engine, No. 2, remains stationary, act upon it with the full power belonging to its density or elasticity, and will thereby cause it to commence its motion downwards; but as the piston of No. 2 moves downwards the density and elastic force of the steam will diminish in proportion as the spaces which it occupies are increased; so that (in case the cylinders of the two engines are of an equal capacity) when the piston of No. 2 has arrived at the bottom or lower end of its stroke the density and elastic force of the steam will be only one-half of what they were while the piston remained at the top. Therefore, if a simple lever, wheel, or working beam, is used for this secondary engine, No. 2, the engine ought only to be loaded with a column of water, or other work, equal to half the number of pounds on each square inch which the primary engine, No. 1, is able to work with; but if the secondary engine, No. 2, is furnished with any proper contrivance for equalizing the power of the steam, it may, in case the cylinders of the two engines are of equal capacity, be made to do seven-tenths of the work which is done by the primary engine, No. 1. When the piston of the secondary engine, No. 2, has come to the bottom of its stroke, the middle regulating valve O is opened, and the steam rushes into the condenser GK, and in its way meets the jet of injection water, which condenses it; and thereby the upper part of the cylinder of the secondary engine, and the lower part of the cylinder of the primary engine, are exhausted of steam. The piston of the secondary engine, No. 2, having then a vacuum both above it and below it, is pulled up easily by the working beam of that engine; and, there being a vacuum under the piston of the primary engine, No. 1, the steam from the boiler exerts its power upon it, and presses it down, and the other motions are repeated, as has been described. These compound engines may also be wrought in other manners, of which I shall describe one of the best. Let the eduction pipe N be supposed to be removed, and a steam pipe S, (which is delineated in red ink), be made to communicate between the perpendicular steam pipe C of the primary engine, and the top regulator box or cross pipe Q of the secondary engine; then the piston of the primary engine, No. 1, being pressed to the bottom by steam, shuts its top regulating valve D, and opens the top regulating valve Q of the secondary engine: the piston of that engine will immediately begin to descend with a decreasing power, (as has been remarked before). When the piston of the secondary engine, No. 2, has come to the bottom of its stroke, its middle regulator O is opened, whereby the steam rushes out of the cylinders of both the engines into the condenser or condensers; and, there being vacuum both above and below the pistons of both engines, the equilibrium of both is restored, and both the pistons are permitted to be raised by the unbalanced weight of the pump rods, or other weights or machinery applied for that purpose. It is proper, in this mode of application, to make a small pipe leading from the lower part of the cylinder of the primary engine to the eduction pipe, or condenser, of the secondary engine; whereby the vacuum under and above both pistons may be maintained of an equal degree of rareness or perfection. For the more clear understanding of these improvements and contrivances I have delineated them on the parchments hereunto annexed, according to scales specified on the respective drawings, and have adapted them all, except figure 1st, to engines whose cylinders are thirty inches in diameter, and the whole length of the stroke of whose pistons is eight feet; but I make the cylinders larger or lesser, longer or shorter, and vary the proportions and shape of them and of the other parts, according as their uses may require; and as each improvement, method, piece of mechanism, or contrivance, admits of numberless variations, I have set forth and delineated only such as I esteem to be among the best, and which are the most easy to be executed.

My Fourth new Improvement on steam or fire engines consists in applying a certain mechanical contrivance, called a toothed rack and sector of a circle, or toothed racks and toothed sectors of circles, for suspending or connecting the pump rods or pistons with the working beams, levers, or other machinery used in place of them, in place or instead of chains, which have hitherto been used for these purposes. This new improvement, or mechanical contrivance, is delineated at OQ, figure 12th, and requires no other explanation than to say that it is delineated by a scale of one-fourth of an inch to each foot of the real size, according to its proper dimensions for a cylinder thirty inches in diameter; and the said rack and sector are supposed to be made of hammered or cast iron, but it may be made of wood or other materials by giving it dimensions suitable to the strength of the material of which it is made; and, in order to accommodate the same to cylinders of other sizes, the strength of its parts must be increased or diminished, in proportion to the powers of the respective cylinders to which it may be applied. I have described and delineated all my aforesaid new improvements upon, and new mechanical contrivance applicable to steam engines, as applied to or connected with the new steam engines of my own invention, as being the most perfect hitherto made; BUT IT MUST BE REMEMBERED, that I do apply the same to the common steam engines known by the name of Newcomen's Steam or Fire Engines; and that they are also applicable to any other species or variety of steam engines which works with a piston moving in a cylinder or steam vessel; and that they will in such engines produce greater or lesser effects in proportion to the degree of perfection of the engine to which they are applied. It is also to be remarked, and though I have described all the engines standing erect, and having the piston rods coming through holes in the top of the cylinder, that the same or nearly the same effects will be produced, though the piston rods come through holes in the bottom of the cylinder, and the working beams or equalizing machinery be placed under them; or although the cylinders and working beams are placed inclined or horizontally; and that, in certain cases, I use them in such different positions.

MY FIFTH NEW IMPROVEMENT on steam or fire engines consists in making the steam vessels in the form of hollow cylinders, or in the form of other regular round hollow vessels, or in the form of greater or lesser segments or sectors of such bodies or vessels. And I place in the centre or axis of the circular curvature of such vessels, a round shaft or axle, passing through and extending beyond one or both ends of such steam vessel; and I shut up the ends of the said steam vessel with smooth plates, which have proper apertures for the said axle or shaft to pass through; and within the said steam vessel I fix to the said axle a piston or plate, extending from the said axle to the circular circumference of the steam vessel, and also extending from one end of the steam vessel to the other. And I make the said piston steam-tight, by surrounding the parts which fit to the steam vessel with hemp, or other soft substances, soaked in grease, oil, or wax, or by means of springs made of steel, or other solid and elastic or pliable materials; and to the said steam vessel I fix one or more plates or divisions, extending from the axle to the circumference of the steam vessel; and, where these plates or divisions join to or approach the axle, or where the said axis passes through the end plates of the steam vessel, I make such joinings steam and air-tight, by the means above recited. In the steam vessel, on each side of the said piston, I make one or more channels or apertures for receiving and discharging the steam, which channels I furnish with proper valves for that purpose: I also apply to the said engine proper condensers and air pumps; and the pumps which raise water, or such other machinery as is required to be wrought by the said engine, are put in motion or worked by a wheel or wheels fixed to or upon the external parts of the said axle, or by any other mechanism which may be suitable. And the said engine, so constructed, is wrought by admitting the steam between the fixed division and the moveable piston, and exhausting or making a vacuum on the other side of the said piston, which accordingly, by the force of the steam, moves into the said vacuum, and turns the axle a greater or lesser portion of a circle, according to the structure of the machine. The piston is returned to its former situation by admitting steam on the other side of the said piston, and drawing the piston back by some external power, or by exhausting the part of the steam vessel which was first filled with steam. An engine constructed upon this principle is delineated at figures 15th, 16th, and 17th in the annexed drawing, according to a scale of one-fourth of an inch to each foot of the real size; but I make them greater or lesser, and vary the shape and size of the steam vessel and other parts, according to their use. The operation of the engine is as follows: The steam vessel AAA being exhausted of steam and air, and the regulating valves K and J being shut, and L and H open, the steam coming from the boiler through the pipe M enters the steam vessel by L and G, and causes the piston C to turn round into or towards the exhausted part of the steam vessel AX, and thereby turns the axle B, and the machinery attached to it, until the piston C comes to X: the regulating valves L and H are then shut, and K and J are opened: the steam which had entered by the pipe G, and had acted upon the piston, returns through G and J into the condenser or eduction pipe O, where it is condensed; and the steam from the boiler, entering through K and F, forces the piston C to return to its first situation. The pump rods UW, or other machinery, are wrought by the wheel SS fixed on the axle BB, or otherwise, (see figures 16th and 17th); and the condenser pump (or pumps) is wrought by the wheel Q, fixed to any part of the said axle, or otherwise. It is to be understood that the said steam vessel is to be firmly fixed, so that it cannot move; and that the gudgeons or pivots of the said axle must be rested upon proper supports, which things could not be exhibited in the drawing without confusion. I also cause engines, made according to this Fifth improvement, to revolve with a continued rotative motion, by making their steam vessels complete cylinders, or other circular figures; and, in place of the fixed division or divisions, I place one or more valves in their steam vessels, which shut or close the area between their axles and their circumferences, and which valves open by turning upon a hinge or joint, or are drawn back by a sliding motion like a drawer, or otherwise are constructed so that they may be removed when the piston comes to them, and thereby suffer it to pass by the place where they were, (see the drawing, figure 18th), and so begin a new revolution in the same direction; or I make a fixed division, or divisions, as has been described, and I fix one or more valves to the axle, which valves are capable of folding down, and of applying themselves to the axle, and of forming a part of its circumference, so that they can thereby pass by the division; and, when they have passed it, they are raised up by springs, or otherwise, so as to perform the office of a piston or pistons. I have not described the boilers which supply any or all of these engines with steam, because I use such as are commonly applied to other steam engines, or any kind of boiler which is capable of producing steam in sufficient quantities. Neither have I described the machinery which opens and shuts the regulating valves, as it is similar to that which is in common use, and may be varied at pleasure.

IN WITNESS whereof I have hereunto set my hand and seal this third day of July, in the year of our Lord one thousand seven hundred and eighty-two.

JAMES WATT.

Sealed and delivered, being first duly stamped, in the presence of
JOHN SOUTHERN.
PETER CAPPER.

Acknowledged by the said James Watt, party hereto, the third day of July, one thousand seven hundred and eighty-two, before me,
GEO. HOLLINGTON BARKER,
A Master in Chancery Extraordinary.

INROLLED in His Majesty's High Court of Chancery, the fourth day of July, in the year of our Lord 1782, being first duly stamped according to the tenor of the Statutes made in the sixth year of the reign of their late Majesties King William and Queen Mary, and in the seventeenth year of the reign of His Majesty King George the Third.
JOHN MITFORD.

4. P. Y.

27. James Watt to James Watt junior,
12 November 1808, Doldowlod.

It is clear from the Boulton and Watt papers that most of the inventions which Watt patented in 1781–82 [23, 26], as well as those later patented in 1784 [29], had been thought of and experimented on by him several years before they were actually patented, together with others which he never patented. Hence the piratical imitations by such men as Wasbrough, Pickard, and the Hornblowers. The following letters [27,28] are a sample from this evidence, referring to his earliest applications of such ideas as the blowing valve, the expansive principle, the crank, and parallel motion.

'Dear James

'I cannot recall whether the blowing valve was used first at Tingtang Engine or not. It was however no invention of Jabez Hornblower being an essential part of Newcomens Engines and perfectly well known to me. If it [had] been first added there it must have been by my direction or consent, as that Engine was erected by my directives and under my superintendance though Mr. Hornblower was the workman employed to put it together and if not employed by B[oulton] and W[att] at least was employed by their approbation and consent for on that point they had an absolute power. We did employ a brake to exhaust the engine to know if the joints were tight and also to set it to work for reasons I do not now recollect and which probably from alteration of construction have ceased to exist. But I certainly was not ignorant of the use of a blowing valve, nor is it supposable though I might have reasons for not using it.

'The great aim of Hornblowers arguments seems to be that the first engines we erected in Cornwall were not so perfect as those we made afterwards, which is perfectly true, the *mechanism* of the engine was not invented all at once, but has been in a course of improvement even unto this day. I was ignorant of many things I know now and possibly of many things which were known to others. Nor do I pretend that I ever possessed an eminently inventive genius in *Mechanicks*. my forte seems to have been reflection and judgement in combining and applying things invented by my self and others. He finds fault with many of the contrivances described in my patents without considering that it was my business not only to describe the things I preferred but also those by which it could be worked, and that patents can only contain first ideas, the very nature of them being that they are new inventions not yet reduced into extensive practice which alone can give perfection. But to judge of the merits of the mechanism of my first engines in Cornwall they were to be compared as they were, with those made [by] his father his brother and himself then at work in the county which they superseded, even in the eyes of Cornish men, no very partial judges towards me.

'I know nothing of Mr. John Dryden or his motions. If I was not the inventor of the paralel motion, I know not who was: for the triangle in Suardi only goes to show the possibility of describing a straight line by composition of circular motions and is otherwise upon a totaly different construction.

'A year or more previous to Jonathan Hornblowers patent an Engine on the expansive principle was at work at Soho, that is an Engine in which the piston ascended in vacuo and in which the steam was admitted only during a portion of its descent and acted during the remainder of the stroke by its elasticity, but in general we made them to ascend in steam on both sides the piston which was the reason for the explanatory patent in 1782.

'In respect to Hornblower his mechanical abilities may be judged of by his engine at Meux's and others where he was under no contract, taking first away what was mine, and his prudence and industry by his being now a wretched prisoner in the Kings Bench almost starving. Whatever is said in reply to his aspersions I wish nothing to be said ad hominem. There is no use in perpetuating enmity and he cannot be more despicable than he is, and the same reasoning will apply to Arthur Wolfe and others of that genus, as to their coadjutor Gregory, it may be proper to give him such answers as may lessen his readers credence in his veracity and discrimination, so as if possible to lessen the evil his books may do, and if a number or two of his pantologia were reviewed it might possibly lessen the number of his dupes.

'I have got a bad headache to day, but shall see Lawson to day or to morrow and shall enquire if he or Murdock know any thing on the subjects you mention.

I remain,

Dear James

Your's etc.

J. WATT'

28. James Lawson and William Murdock to James Watt junior, *14 November 1808, Doldowlod.*

'Dear Sir

'In answer to your inquirys – of the 8th inst. – The exact time at which the blowing Valve was put to Engines is unknown. Mr. Murdock however perfectly recolects one being put on the pumping Engine here in the end of the year 1777 – He having assisted Joseph Harrison in putting it on – Mr. M[urdock] has good reason to remember it from being scalded by the Engine man's blowing through, while he was in the Cistern, for at this time the Valve was put inside the condensing cistern – it was soon afterwards taken outside to a small cistern as at present – There was also a brake used at this time to exhaust the condenser and prove the joints. At the time I came here in April 1779 the blowing Valve was in general use – as at present – Mr. Murdock put one to the Engine on Bedworth Colliery in September 1778.

'We are totally unacquainted with Mr. Dryden – or his parallel Motions.

'Mr. Murdock perfectly recolects Mr. W[att] being busy in drawings and scheming various methods for Parallel Motions, when in Cornwall in the year 1781. The first sort was called the perpendicular Motion – one of which Mr. Murdock put up at Crain Audit in Cornwall about the year 1783 – and the next were at North Downs in 1784 or 5 with Parallel Motions same as at present.

'The expansive principle was used at Soho Engine in October 1777 at which time Mr. M[urdock] saw it and made patterns for New Nozells on the principle when the Engine was enlarged. Mr. M[urdock] also made patterns of expansive Nozels both for Halamanning and Wanlockhead Engines in the year 1778.

'Mr. M[urdock] saw Mr. Watt's model of an Engine working two Cranks at Soho sometime in the year 1779 – (About September) – We both have often heard that Richard had described this Model to the Workmen employ'd at an Engine then working at Birmingham in which a rotative motion was produced by rack work attached to the working beam, R[ichard] C[artwright] having afterwards confessed giving this information to these Workmen – in which he pretended to think there was no harm.

I remain Dear Sir

Yours truly

JAMES LAWSON

I confirm the Contents

WILLIAM MURDOCK

'NB. a Blowing Valve was not absolutely necessary as blowing through the air pump was before practiced – which answered the purpose.

'The first account we can find of [a] Blow valve (amongst the old drawings) is in July 1777 for an Engine at Snedshill and there is holes marked (in the drawings of Ting Tang) in the Eduction pipe – of 3 and $3\frac{1}{2}$ Inch Diameter – in the usual place for that purpose – tho' without any explanation.

'I inclose a letter from old Mr. Hornblower to shew he was to put up the Engine at Ting Tang by the direction of Your Father.'

29. 1784 Specification of Patent

This patent was granted on 28 April 1784 and the specification was enrolled on 24 August 1784. It includes a number of improvements, of major and minor significance: (i) another steam wheel; (ii) various mechanisms, including the parallel motion, for connecting the piston-rod to the beam; (iii) various methods of balancing pump-rods; (iv) new methods of applying steam power to rolling and slitting mills; (v) steam-powered forge hammers; (vi) improved method of opening and closing regulating valves; (vii) application of steam engines 'to give motion to wheel carriages', *i.e.* steam locomotives. Of these inventions, the one of most immediate practical significance was the parallel motion, for connecting the piston-rod of the double-acting engine to the beam, so as to permit 'push' as well as 'pull'. The application of the steam engine to rolling and slitting mills and to forge hammers soon had important effects in ironworks, such as those of John Wilkinson. Steam locomotion, of course, was ultimately to be of the most revolutionary importance, but though William Murdock, Boulton and Watt's engine-erector and foreman, produced a model steam-carriage about this time, he was given no encouragement by Watt, who apparently patented this idea only to forestall other possible projectors.

SPECIFICATION OF PATENT, APRIL 28TH, 1784, FOR CERTAIN NEW IMPROVE-MENTS UPON FIRE AND STEAM ENGINES, AND UPON MACHINES WORKED OR MOVED BY THE SAME.

TO ALL TO WHOM these presents shall come, I, JAMES WATT, of Birmingham, in the county of Warwick, Engineer, send greeting.

WHEREAS His Most Excellent Majesty King George the Third, by His Letters Patent bearing date at Westminster, the twenty-eighth day of April, in the twenty-fourth year of his reign, did give and grant unto me, the said JAMES WATT, his especial licence, full power, sole privilege, and authority, that I, the said JAMES WATT, my executors, administrators, and assigns, should, and lawfully might, during the term of years therein expressed, make, use, exercise, and vend, within that part of his Majesty's Kingdom of Great Britain called England, his Dominion of Wales, and Town of Berwick upon Tweed, my Invention of 'CERTAIN NEW IMPROVEMENTS UPON FIRE AND STEAM ENGINES AND UPON MAC-'HINES WORKED OR MOVED BY THE SAME;' in which said recited Letters Patent is contained a Proviso obliging me, the said JAMES WATT, by writing under my hand and seal, to cause a particular description of the nature of the said invention, and in what manner the same is to be performed, to be inrolled in His Majesty's High Court of Chancery within four calendar months after the date of the said recited Letters Patent, as in and by the said recited Letters Patent, and the Statute in that behalf made, relation being thereunto had, may more at large appear.

NOW KNOW YE, That in compliance with the said Proviso, and in pursuance of the said Statute, I the said JAMES WATT do hereby declare, that the nature of my said invention, and in what manner the same is to be performed, is particularly described and ascertained as follows, that is to say:

MY FIRST NEW IMPROVEMENT on steam and fire engines consists in making the steam vessel so as to be capable of turning round on pivots or on an axis, either in a vertical or horizontal direction, and in employing the elastic power of the steam to press upon the surface of any dense fluid or liquid contained in the steam vessel, and to force it to pass out at a hole or holes made in the circumference or external part of the steam vessel, in such manner that the fluid or liquid shall issue out in a line forming a tangent to the circle described by the rotation of that part of the steam vessel where the hole is situated, or at least in a line approaching to such tangent, which fluid or liquid, by its action on the fluid or liquid in which the steam vessel is immersed, causes the engine to turn round; and I replace the fluid so issuing, by immersing the steam vessel into another vessel filled with the same fluid, and by dividing it into two or more divisions or chambers, in the direction of its axis of rotation; which chambers are furnished with valves in their bottom or sides, which admit the fluid to enter into one of them, while by the force of the steam it issues from another, and the steam is admitted into these chambers, and its action is suspended alternately by proper valves and regulators, which are opened and shut by the rotation of the machine. The machine may be made in various forms: I have delineated one of the most commodious in figure 1st of the annexed drawing, where ABCDE represents a section of the steam vessel mounted on a vertical axis, with one perpendicular partition; F, G, two valves in its bottom, to admit the surrounding liquid; H, the place of one of the holes at which it issues; J, the pivot on which it turns; K, a collar which is fitted exactly to the neck of the steam vessel, so as to be steam-tight, and yet to give liberty to the vessel to turn round in it freely; L, a pipe conveying steam from a boiler to the steam vessel, and which is joined to the collar. There are two holes in the neck of the steam vessel which communicate with its two chambers, and, as the vessel turns round, present themselves alternately to the opening of the steam pipe within the collar, and to an opening on the opposite side of the collar, which communicates with the empty part of the vessel MN, which contains the liquid in which the steam vessel is immersed. The vessel MN is made of a cylindric or other form, and contains a quantity of quicksilver, water, oil, or other fluid or liquid, of a depth sufficient to give a resistance to the issuing fluid proportioned to the effect wanted to be produced, and to cause the fluid to fill the empty chamber of the steam vessel in proper time. If the machine is required to act only by such steam as has a greater expansive force than the pressure of the atmosphere, this vessel MN is opened at top, and the steam which has done its office is then discharged into the open air; but when the steam is required to act with the pressure of the atmosphere, in addition to its own elastic force, then the vessel MN must be shut at top, and made everywhere air-tight, and the steam be made to pass from it by the pipe O into another vessel, called a condenser, where it is condensed by the application of cold water, or other cold bodies, by which means a vacuum is obtained, and the power of the steam is augmented. Figure 2nd is a ground plan of the steam vessel, and of the external vessel; HH are two apertures at which the fluid alternately issues, and FG are the valves by which it enters. Figure 3rd is a horizontal section of the collar and of the neck of the steam vessel, in which RR are the two apertures which admit and discharge the steam, and L the steam pipe coming from the boiler. The action of the machine is as follows: Steam being admitted into one of the chambers, causes the fluid contained in it to issue with violence at the lower aperture belonging to that chamber, and the fluid so issuing, by its action against the quiescent fluid contained in the outer vessel, causes the steam vessel to turn round; by which means the upper aperture of the other chamber is presented to the steam pipe, and the action on the fluid in that chamber commences; at the same time the steam issues at the upper aperture of the first chamber, and it is replenished by the fluid entering by the valve at the bottom. The machinery to be worked by the engine is put in motion by being attached, fixed to, or connected with the end of the axis P, which passes through a hole or collar in the cover of the outer vessel MN. This engine is delineated by a scale of one inch for each foot of the real size, but may be larger or lesser, and its proportions varied at pleasure.

MY SECOND NEW IMPROVEMENT on the steam engines consists in methods of directing the piston rods, the pump rods, and other parts of these engines, so as to move in perpendicular or other straight or right lines without using the great chains and arches commonly fixed to the working beams of the engines for that purpose, and so as to enable the engine to act on the working beams or great levers, both by pushing and by drawing; or both in the ascent and descent of their pistons. I EXECUTE THIS ON THREE PRINCIPLES: THE FIRST PRINCIPLE is delineated in figure 4th, and is performed by connecting with the top of the piston rod A by means of joints, two secondary levers or beams BC, the other ends of which are furnished with arches DE, which roll upon the walls of the house FG, or on pieces of timber or other proper resisting bodies fixed at convenient distances parallel to the line of motion of the rod which is required to receive the right-lined or perpendicular motion; and these arches are suspended in their places by means of leather belts, chains, or jointed bars of iron, attached to them. The top of the piston or pump rod, and the working beam, are also connected together by means of a piece of iron or wood H, having a joint at each end, on which it can move freely and accommodate itself both to the angular motion of the working beam JJ and to the perpendicular motion of the piston or pump rod. THE SECOND PRINCIPLE by which I produce a right-lined motion from an angular or circular motion is delineated in figures 5th and 6th, and consists in guiding the top of the piston rod or pump rod perpendicularly by means of a piece of wood or iron, or other material A, sliding in a groove made in an upright B, firmly fixed to some part of the machine, so as to be higher than the working beam C, and in the direction of the required motion, and by connecting the end of the working beam with the top of the piston rod or pump rod D, or with the said sliding piece A, by means of a bar or bars of wood or iron E, having joints at each end. Figure 6th shows an end view of the working beam, piston rod, sliding piece, and connecting bars; and the pricked circular and angular lines, figure 5th, show the quantity of angular motion of the working beam. In figure 19th is delineated another method of using this principle: AA is the piston of the engine, which has a hollow piston rod BB, and the connecting bar CC is attached to the bottom of that hollow rod by means of a joint, and to the working beam by means of another joint, and the hollow piston rod, by sliding in the collar DD, serves to direct the motion, and the hollow rod is made wide enough to permit an angular motion of the connecting bar, proportioned to the versed sine of the angle described by the working beam. THE THIRD PRINCIPLE upon which I derive a perpendicular or right-lined motion from a circular or angular motion consists in forming certain combinations of levers moving upon centres, wherein the deviation from straight lines of the moving end of some of these levers is compensated by similar deviations, but in the opposite directions, of one end of other levers. In figures 7th and 8th is delineated one Method of putting this principle into practice: AA

114

represents the working beam of the engine; B, the wall or support on which it rests; CC, the spring beam; DE, a lever or bar of iron, moveable about the axis or centre D, (which is fixed to the spring beam), and also about the axis or centre E, which is connected with, or is a part of, the bar or rod EF. The bar EF is capable of sliding towards or from the axis or centre of the working beam by a sliding motion between a pair of cheeks at E, and by a portion of a circular motion of the end F round the centre H, from which it is suspended by the coupling bars or links HF. The piston rod or pump rod G is fastened to the bar FE and to the link HF by the joint pin at F. The right line FL is that in which the piston rod or its end F moves upwards and downwards. KM is a portion of a circle described from the centre of the working beam to which LF is a tangent, and KL and MF show its greatest deviations from the straight line or tangent. ENJ is a portion of a circle described on the centre D by the joint pin E, and the length of DE and the situation of the centre D are such that the lines JL, NO, EF, are equal, and are represented by the rod EF; therefore, whenever the working beam is moved on its centre, the cheeks at E and the joint at F oblige the bar EF to move upwards or downwards along with it. But the centre D being fixed, the end E of the rod ED can only move in the circle ENJ, and the joint pin E obliges the rod EF to slide under the working beam between the cheeks, and to come nearer to the centre or to recede from it, according to the position of the working beam. And as the distances between every point of the arc JNE and the corresponding points of the line LOF are always equal, the point or joint F, together with the top of the piston rod, must move in the straight line LOF. In figure 8th is delineated a view of the underside of the working beam, seen from below, in which it is shown that the bar ED is made double, in order that one part may go on each side of the working beam, and that the joint pins E, D, and F, reach from side to side of the beam. Though the apparatus is in these drawings placed below the working beam, and that beam is fixed above its own axis or gudgeon, yet the contrivance will answer equally well if the whole were reversed, or if the rod EF and the axis of the working beam were placed in the middle of its depth by making the beam double. The proportions of the lengths and thicknesses of the parts to one another are such as answer in practice, but may be considerably varied, provided the principle be attended to. In figures 9th, 10th, and 11th, I have delineated another Method, whereby I carry this third principle into practice. AA represents the working beam of an engine; BD, the piston rod or pump rod; CDE, one of two bars of iron or wood, connected, by joints at E and D, with the working beam and with the top of the piston rod respectively, and at C, by a joint, to an angling bar or bars CF, the other end or ends of which is or are connected with the wall of the house, or some firm support, by a joint or joints at F, on which, as a centre, the bar CF is moveable. When the working beam is put in motion on its axis, the point E describes the arc HEJ, and the point C describes the arc KCL on the centre F, and the convexities of these arcs, lying in opposite directions, compensate for each other's variations from a straight line; so that the joint D at the top of the piston rod or pump rod, which lies between these convexities, ascends and descends in a perpendicular or straight line. The respective lengths of the radii GE and CF and their proportions to one another may be varied, but if the radius CF be lengthened more in proportion than GE, the point D must be placed pro-

115

portionably further from E and nearer to C, and vice versâ, as is pointed out by geometry. The regulating radius or rod CF may also be placed above the working beam, and the latter may be reversed in regard to its own axis, where such construction is found more convenient. In figure 10th I have delineated a horizontal view of the radii CF, and in figure 11th an end view of the working beam A and the connecting bars EDC with the piston rod BB. In figure 12th I have delineated another Method, whereby I put this third principle into practice. The regulating radius FH being centred at H on a firm support, and the connecting bar FBG being connected with the working beam by a joint at B, is prolonged to G, where it is connected with the piston rod or pump rod GC by another joint. As the radius FH is shorter than the radius JB, the parts of the connecting rod FB, BG are so proportioned to one another, that on account of the difference of the convexities of the arches FE and BC, the point G and the piston rod GC will always ascend and descend in a perpendicular or straight line. This machinery also admits of being reversed; that is, the regulating radius HF may be placed under the working beam, and the working beam above its own centre of motion, in cases where that is found more convenient. This third principle may also be put in practice by other methods, but those delineated are in general the most eligible: all of them are laid down by a scale of one-fourth of an inch for each foot of the real size, in their proper proportions for engines whose cylinders are twenty inches in diameter, and the length of whose stroke is four feet; except figures 5th and 6th, which are delineated for a six feet stroke; and figure 19th, which is laid down for a cylinder fifteen inches in diameter and twelve inches length of stroke; but all the dimensions admit of considerable variation, according to the exigency of the case, and, preserving the proportions, are applied to cylinders of different diameters and lengths of stroke.

MY THIRD NEW IMPROVEMENT is upon the application of steam or fire engines to work pumps or other alternating machinery. It consists in causing, by proper machinery, one half or one part of the pump rods to ascend while the other descends, whereby the weight of the pump rods, and other moving parts, always nearly balance one another, and the necessity of employing other matters to balance their weights, which are frequently enormous, is avoided. It is particularly applicable to engines which act forcibly both by the ascent and descent of the piston in the cylinder, but may also be advantageously employed to engines which act forcibly by the motion of the piston in one direction, as in such cases the weight of one set of pump rods may be employed, during their descent, to pull up the rods of other pumps, and thereby to work them: I PERFORM THIS IN VARIOUS METHODS, of which I have delineated two of the best. THE FIRST OF THESE consists in suspending one-half, or part of the pump rods, to one end of a lever or working beam; and the other part, or half, to the other end of the same lever, which lever or working beam is supported on a centre or axis in some part of its length between the two rods, and the said rods are connected with or suspended to or from the said double lever by means of simple joints or by means of chains and arches, or, in place of a lever, a wheel and chain may be used. The said lever may either be the working beam of the engine, or a separate lever placed over, or within, the mouth of the pit or shaft, and receiving its motion either directly from the piston rod of the engine, or from the working beam by means of a stiff rod; or, in case of the engine being of that kind which acts only in one direction, by means of a chain or rope. In figure 13th, AA represents the double lever; B, its centre or axis; CL and DM, two rods of wood or iron, or two chains connecting this beam with the pump rods, which pump rods are suspended to the joints ML. KK represent two wheels which serve to guide the rods, and to bring them nearer together than they would naturally hang, in cases where the connecting rods CL, DM cannot be admitted to hang perpendicularly under the ends of the double lever; and J represents the lower end of a stiff rod, reaching from the working beam, or other moving part of the engine, and connected with the double lever AA by means of the joint H. AND BE IT REMARKED, that in cases where the connecting rods CL, DM are kept from their perpendicular position by means of wheels, (such as KK), the lower ends ML of these rods would describe curved or angular lines, which would disturb the perpendicularity of the pump rods suspended from these points. In order, therefore, to avoid such defect, I form the lower part of these connecting rods into curves, as shown in the drawing, which curves, acting against the wheels KK, compensate for the irregularities which would otherwise take place, and cause the points L and M, and the pump rods which are attached to them, to ascend and descend in perpendicular lines: (the figure of such curve is readily found by geometry). THE SECOND METHOD consists in connecting together levers which turn or move on separate or different axes or centres, in such a manner, that the moveable end of one lever shall descend when the moveable end of the other lever ascends, and vice versâ; and in connecting the moveable end of one of these levers with the working beam or piston of the engine in some proper manner, and in suspending the pump rods to the moveable ends of these levers by means of joints or chains. No. 14 represents a combination of levers which comes under that description. The two similar frames BFEH and CGDJ, are moveable on their respective centres E and D, and are connected together by a brace or braces HKJ, so that the pieces HE and JD must, when put in motion, continue parallel to one another; and consequently, from the positions of the pieces or levers F and G, when the joint pin B describes the arc BN, the joint pin C of the other frame must describe the arc CO in the opposite direction, by which means the pump rod L ascends when the pump rod M descends, and vice versâ; and these frames are connected with the engine, or its working beam, by means of the rod or chain AB; or in place thereof, one of the levers F or G is prolonged on the other side of their respective centres E or D to a sufficient length to become itself the working beam of the engine. And the whole of this machinery may be reversed or placed upside down: that is, the levers EH and DJ, with the brace or braces HKJ, may be placed above their centres E and D instead of being placed below them, and may also upon the same principle be varied in other manners. The machinery delineated in figures 13th and 14th is laid down by a scale of one-fourth of an inch for every foot of the real dimensions, and is adapted for cylinders of twenty inches diameter and four feet stroke; but machinery on the same principles is made for larger or lesser engines, as required.

MY FOURTH NEW IMPROVEMENT consists in new methods of applying the power of steam engines to move mills for rolling and slitting iron and other metals, or to move other mills which have many wheels which are required to turn round in concert, so that the same steam or fire engine shall directly, by means of a double working beam, or by means of a strong piece of wood or other material fixed across one end of the working beam, and by means of two separate rods connecting the said working beam or cross beam with proper machinery for producing rotative motions, give motion to two primary wheels fixed either on the same or separate axes, whether acting in concert or applied to different uses; and in connecting together, by means of a secondary axis carrying two or more wheels, different primary motions produced by the same engine, or by two or more different engines, which methods are particularly applicable to the connecting together the motions of the rollers and slitters, or of different pairs of rollers, in mills for rolling and slitting metals, which are worked by fire or steam engines. In figures 15th, 16th, and 17th is delineated the application of this improvement to a slitting mill, which drawings exhibit that machine in three different views, in all of which the same parts are marked with the same letters. Figure 16th is a section of the mill through the line XX of figure 17th, except in what relates to the working beam AB, the cross beam C, and the upper ends DE of the two connecting rods of the rotative motions. In order to avoid confusion, the rod E is represented as broken off, and the remaining part of it and the position of the rotative motion which it works are represented by the red ink lines and circles Q and R. F is the revolving wheel of a rotative motion, which is fixed to the rod D, and GG is the centre wheel of the same motion, which is fixed upon the axis of the fly wheel JJ and of the toothed wheel KK, to both which it gives motion, and also to the lower set of the slitters which are turned by that axis. The toothed wheel KK acts upon and gives motion to another toothed wheel NN, which is fixed on the same axis with the fly wheel OO, which turns the upper roller in the roll frame P, and the direction of the motion of these and the other wheels is shown by the darts. And the rod E, with the rotative motion QR, (figures 15th and 17th), turns the fly wheel MM, and the toothed wheel LL, which is upon the same axis with that fly, and which axis also turns the upper set of the slitters. The toothed wheel LL acts upon and turns the toothed wheel SS and its fly wheel TT, and their axis turns the lower roller. The motion is thus communicated to all the wheels and flys, and also to both the rollers and slitters. But as the flys JJ and TT must turn in the opposite direction to the flys MM and OO, and as there is nothing in the rotative motions themselves that can determine these wheels to take contrary motions at first setting out, and as the top and bottom of the motion of the connecting rods D and E, and the revolving wheels which are fixed to them, the necessary shake in the teeth of the wheels of the rotative motions will permit the fly wheels on one side of the mill to run faster than those on the other, which produces a very prejudicial effect on the whole machine, I have contrived to connect the toothed wheels KK and SS together, by means of the toothed wheels V and W, fixed on the secondary axis Y under the other machinery: the teeth of the wheel V are engaged into the teeth of the wheel K, and the teeth of the wheel W are engaged into those of the wheel S; so that W and V being on the same axis, cause K and S to turn in the same direction, and consequently L and N to turn in the contrary direction, and the whole wheels to move in an uniform manner. In place of the cross beam C, two working beams may be used, united together at the ends which are next the cylinder, and

118

opening to the proper distance at the other ends like the letter V, and one of the connecting rods may be suspended to one of these beams, and the other rod to the other beam. In other cases, where the power required is not very great, I construct slitting mills with engines which have only one working beam and one set of machinery for producing a rotative motion, which is connected with the axis of one of the fly wheels, from whence the motion is transmitted to the other parts of the machines by the toothed wheels KK, LL, NN, and SS, and by the secondary axis Y and its toothed wheels V and W. The whole of this machinery is drawn in its due proportions, to a scale of one-fourth of an inch to each foot of the real size, to be moved by an engine with a cylinder of forty-eight inches diameter and six feet length of stroke, but it is made larger or lesser, as required. BE IT REMEMBERED, that although in order to explain my said Fourth new improvement, I have been obliged to delineate the whole of a slitting mill, yet that the new improvements which are the subjects of this article of the Specification, are only, the communicating the motion from the same steam or fire engines to two separate primary axes or shafts and sets of wheels moving in the same or in contrary directions; and in connecting together four or more sets of wheels by means of secondary axes or shafts carrying two or more wheels each, such as the shaft or axis Y and its wheels V and W; both which improvements are, to the best of my knowledge, entirely new. In figure 20th I have delineated the machinery by which the rotative motion is in this case proposed to be communicated from the working beam of the engine to the millwork, (which is one of the methods described in the Specification of certain Letters Patent which his present Majesty was most graciously pleased to grant to me, bearing date the twenty-fifth of October, in the twenty-second year of his reign). But this drawing is only intended for elucidation, for I apply to this purpose not only the rotative machinery now delineated, but also any other kind which is proper or suits the particular case.

MY FIFTH NEW IMPROVEMENT consists in applying the power of steam or fire engines to the moving of heavy hammers or stampers, for forging or stamping iron, copper, and other metals or matters, without the intervention of rotative motions or wheels, by fixing the hammer or stamper, to be so worked, either directly to the piston or piston rod of the engine, or upon or to the working beam of the engine, or by fixing the hammer or stamper upon a secondary lever or helve, and connecting the said lever or helve by means of a strap, or of a strong rod, to or with the working beam of the engine, or to or with its piston or piston rod. In figure 18th is delineated, on a scale of one-third of an inch for each foot of the real size, a view of an engine of my invention, with a cylinder of fifteen inches diameter, working a hammer of five hundred pounds weight: in which, A is the cylinder; B, the piston rod; CC, the working beam of the engine; DD, the drom beam of the forge, which in this case is made double; E, a strong post, which carries the end of the drom beam; F, the rabbit or spring piece which regulates the ascent of the hammer and beats it back; GG, one of the two legs which carry the axis of the hammer; H, the hurst or axis of the hammer helve; J, the helve; K, the puppet or piece which supports the rabbit; L, the hammer; N, the anvil, and M, the rod which connects the helve with the working beam; OP, the nozzles or regulator boxes; R, the condensing vessel;

I

S, the condenser pump; TT, a cistern of cold water; V, the plug frame which opens and shuts the regulators; and W, a steam pipe coming from a boiler. In figure 19th is delineated, on a larger scale, a section of the cylinder, and a view of the apparatus by which, in cases where the engines are required to be worked very quick, the regulators are opened and shut; which shall be described in the next section.

My Sixth new Improvement consists in making the regulating valves which admit the steam into the cylinders of fire or steam engines, or which suffer it to go out of them, in such manner that they are pushed open by the action of the steam upon them, and are kept shut by certain catches or detents, which are unlocked at proper times, either by hand or by the engine itself. I perform this by making the circumference of these valves in a conical or tapering form, as shown at P and Q, figure 19th; and by grinding or otherwise fitting exactly the said circumferences to a ring of metal which is called the valve seat, so that when shut it may be steam or air-tight: the valves Q and P are suspended or supported by links or by racks and sectors which connect them with certain levers E and F in the insides of the nozzles, which, by means of the spindles or axes on which they are fixed, communicate with the levers GG which are upon the outside of the nozzles, and are connected by the rods HH with the short levers ST and VW turning upon their respective axes S and V. When the valve is shut, the centres of the pins T and W lie a little beyond the straight lines, passing through the centres of S and G and of V and G, and no force applied at GG can unlock or discharge the valves P and Q until the points T and W are moved to the other sides of the straight lines SG and VG respectively. When the piston of the engine descends, the valve P is shut and Q is open; the pin X strikes the handle K, which shuts Q, and the pin M immediately strikes the handle J, which, moving the point T from the right line, disengages the valve P, and suffers it to open. The piston AA then ascends with the plug tree L, and the pin Z returns the handle J to its proper position, which shuts P, and the pin N strikes the handle K, which permits Q to open, and the piston AA descends into its first position and commences a new stroke upward, and so continuedly. And the power and velocity of the engine is regulated by opening or shutting a regulating valve placed at O, which admits more or less steam, as required.

My Seventh new Improvement is upon steam engines which are applied to give motion to wheel carriages for removing persons or goods, or other matters, from place to place, and in which cases the engines themselves must be portable. Therefore, for the sake of lightness, I make the outside of the boiler of wood, or of thin metal, strongly secured by hoops, or otherwise, to prevent it from bursting by the strength of the steam; and the fire is contained in a vessel of metal within the boiler, and surrounded entirely by the water to be heated, except at the apertures destined to admit air to the fire, to put in the fuel, and to let out the smoke; which latter two apertures may either be situated opposite to one another in the sides of the boiler, or otherwise, as is found convenient; and the aperture to admit air to the fire may be under the boiler. The form of the boiler is not very essential, but a cylindric or globular form is best calculated to give strength. I use cylindrical steam vessels with pistons, as usual in other steam engines, and I employ the elastic force of steam to give motion to these pistons, and after it has performed its office I discharge it into the atmosphere by a proper regulating valve, or I discharge it into a condensing vessel made air-tight and formed of thin plates or pipes of metal, having their outsides exposed to the wind, or to an artificial current of air produced by a pair of bellows, or by some similar machine wrought by the engine or by the motion of the carriage; which vessel, by cooling and condensing part of the steam, does partly exhaust the steam vessel, and thereby adds to the power of the engine, and also serves to save part of the water of which the steam was composed, and which would otherwise be lost. In some cases I apply to this use engines with two cylinders which act alternately; and in other cases I apply those engines of my invention which act forcibly both in the ascent and descent of their pistons, and by means of the rotative motion in figure 20th, or of any other proper rotative motion, I communicate the power of these engines to the axis or axletree of one or more of the wheels of the carriage, or to another axis connected with the axletree of the carriage by means of toothed wheels; and in order to give more power to the engine when bad roads or steep ascents require it, I fix upon the axletree of the carriage two or more toothed wheels of different diameters, which when at liberty can turn round freely on the said axletree when it is at rest, or remain without turning when it is in motion; but, by means of catches, one of these wheels at a time can be so fixed to the axletree, that the axletree must obey the motion of the toothed wheel, which is so locked to it. And upon the primary axis, which is immediately moved by the engine, or which communicates the motion of the engine to the axletree of the carriage, I fix two or more toothed wheels of greater or lesser diameters than those on the axletree, which are moved by them respectively, so that the wheels on these two axles having their teeth always engaged in one another, the wheels on the axle of the carriage always move with the wheels on the axle of the rotative motion, but have no action to turn the wheels of the carriage except one of them be locked fast to its axletree, – then the latter receives a motion faster or slower than that of the axle of the rotative machinery, according to the respective diameter of the wheels which act upon one another. In other cases, instead of the circulating rotative machinery, I employ toothed racks or sectors of circles worked with reciprocating motions by the engines, and acting upon ratchet wheels fixed on the axles of the carriage. And I steer the carriage, or direct its motion, by altering the angle of inclination of its fore and hind wheels to one another by means of a lever or other machine. As carriages are of many sizes and variously loaded, the engines must be made powerful in proportion. But to drive a carriage containing two persons, will require an engine with a cylinder seven inches in diameter, making sixty strokes per minute of one foot long each, and so constructed as to act both in the ascent and descent of the piston; and the elastic force of the steam in the boiler must occasionally be equal to the supporting a pillar of mercury thirty inches high.

LASTLY, as throughout this Specification I have particularly described my several new Improvements as applied to the improved steam engines of my invention, the sole use and property of which his present Majesty was most graciously pleased to grant to me, my executors, administrators, and assigns, by his Royal Letters Patent, bearing date the fifth day of January, in the ninth year of his reign; and which were confirmed by an

Act of Parliament made and passed in the fifteenth year of his reign; and for sundry improvements on which, his Majesty was also most graciously pleased to grant to me his Royal Letters Patent, bearing dates on the twenty-fifth of October and the twelfth day of March, both in the twenty-second year of his reign; – BE IT THEREFORE REMARKED, the said new improvements herein particularly described are all, or most of them, not only applicable to the improved engines of my invention, but also to engines of other constructions which would be improved thereby; and that, as I suppose the Specifications of explanation of any particulars relative to my former inventions, which may not be clearly understood from these presents.

IN WITNESS whereof, I the said JAMES WATT have hereunto set my hand and seal this twenty-fourth day of August, in the year of our Lord one thousand seven hundred and eighty-four.

JAMES WATT.

Signed and sealed, being first duly stamped, in the presence of
JOHN SOUTHERN.
ZACCHEUS WALKER, junior.

Acknowledged by the within named JAMES WATT this twenty-fourth day of August, in the year of our Lord one thousand seven hundred and eighty-four, before me,
WILLIAM BEDFORD,
Master Extraordinary in Chancery.

INROLLED in His Majesty's High Court of Chancery, the twenty-fifth day of August, in the year of our Lord 1784, being first duly stamped according to the tenor of the Statutes made for that purpose.

4. P. Y.
JOHN MITFORD.

30. 1785 Specification of Patent

This patent was granted on 14 June 1785 and the specification was enrolled on 9 July 1785. The rapid increase in the number of steam engines, together with furnaces of all kinds, began to pollute the atmosphere of industrial areas. In this patent, Watt tried to cope with the problem by means of a smokeless furnace or fire-place.

SPECIFICATION OF PATENT, JUNE 14TH, 1785, FOR CERTAIN NEWLY IMPROVED METHODS OF CONSTRUCTING FURNACES OR FIRE-PLACES FOR HEATING, BOILING, OR EVAPORATING OF WATER AND OTHER

10 Section of fire-engine boiler and its furnace or fire-place, from *Mechanical Inventions of James Watt* by J. P. Muirhead

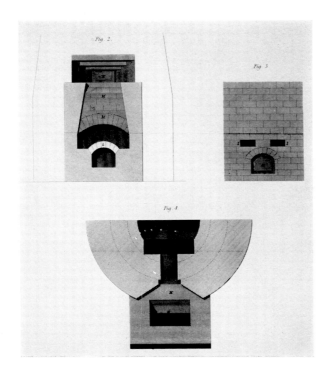

11 *above left:* Section of same fire-place in the other direction, in which MM is the back of the fire-place
above right: Outside view of the same fire-place
below: Plan of fire-place

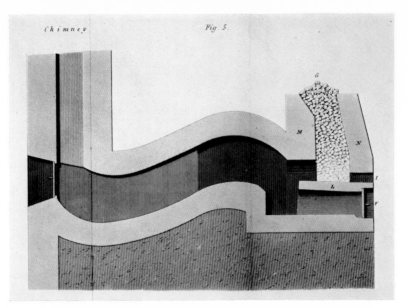

12 The fire-place as applied to a furnace for melting metals

13 The fire-place with additional smaller grate

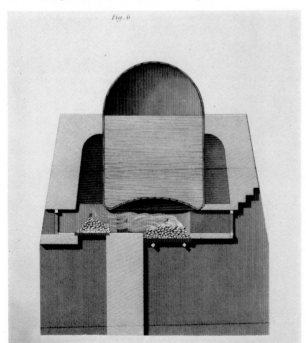

LIQUIDS WHICH ARE APPLICABLE TO STEAM ENGINES AND OTHER
PURPOSES, AND ALSO FOR HEATING, MELTING, AND SMELTING OF
METALS AND THEIR ORES, WHEREBY GREATER EFFECTS ARE PRODUCED
FROM THE FUEL, AND THE SMOKE IS IN A GREAT MEASURE PREVENTED
OR CONSUMED.

TO ALL TO WHOM these presents shall come, I, JAMES WATT, of Birmingham, in the
county of Warwick, Engineer, send greeting.

WHEREAS His Most Excellent Majesty, King George the Third, by His Letters
Patent under the Great Seal of Great Britain, bearing date at Westminster, the fourteenth
day of June, in the twenty-fifth year of his reign, did give and grant to me, the said
JAMES WATT, his especial license, full power, sole privilege, and authority, that I, the
said JAMES WATT, my executors, administrators, and assigns, should, and lawfully
might, during the term of years therein expressed, make, use, exercise, and vend throughout
that part of Great Britain called England, his Dominion of Wales, and Town of Berwick
upon Tweed, my invention of 'CERTAIN NEWLY IMPROVED METHODS OF CONSTRUCT-
'ING FURNACES OR FIRE-PLACES FOR HEATING, BOILING, OR EVAPORATING OF
'WATER AND OTHER LIQUIDS WHICH ARE APPLICABLE TO STEAM ENGINES AND
'OTHER PURPOSES, AND ALSO FOR HEATING, MELTING, AND SMELTING OF
'METALS AND THEIR ORES, WHEREBY GREATER EFFECTS ARE PRODUCED FROM
'THE FUEL, AND THE SMOKE IS IN A GREAT MEASURE PREVENTED OR CON-
'SUMED;' in which said Letters Patent is contained a proviso obliging me, the said
JAMES WATT, particularly to describe and ascertain the nature of my said invention,
and in what manner the same is to be performed, by an instrument in writing under my
hand and seal, and to cause the same to be inrolled in his Majesty's High Court of Chancery
within one calendar month next and immediately after the date of the said Letters Patent,
as in and by the said Letters Patent, and the statute in that behalf made, relation being
thereunto respectively had, may more at large appear.

NOW KNOW YE, that in compliance with the said proviso, and in pursuance of the
said statute, I, the said JAMES WATT, do hereby declare that the following is a particular
description of the nature of my said invention, and in what manner the same is to be
performed (that is to say): My said newly improved methods of constructing furnaces
or fire-places consist in causing the smoke or flame of the fresh fuel, in its way to the flues
or chimney, to pass together with a current of fresh air through, over, or among fuel
which has already ceased to smoke, or which is converted into coke, charcoal, or cinders,
and which is intensely hot, by which means the smoke and grosser parts of the flame,
by coming into close contact with, or by being brought near unto the said intensely hot
fuel, and by being mixed with the current of fresh or unburnt air, are consumed or
converted into heat, or into pure flame free from smoke.

I put this in practice – FIRST, by stopping up every avenue or passage to the chimney
or flues, except such as are left in the interstices of the fuel, by placing the fresh fuel

above, or nearer to the external air, than that which is already converted into coke or charcoal; and by constructing the fire-places in such manner that the flame, and the air which animates the fire, must pass downwards, or laterally or horizontally, through the burning fuel, and pass from the lower part, or internal end or side of the fire-place, to the flues or chimney. In some cases, after the flame has passed through the burning fuel, I cause it to pass through a very hot tunnel, flue, or oven, before it comes to the bottom of the boiler, or to the part of the furnace where it is proposed to melt metal, or perform other office, by which means the smoke is still more effectually consumed. In other cases, I cause the flame to pass immediately from the fire-place into the space under a boiler, or into the bed of a melting or other furnace.

The drawing, figure 1st, shows a section of a fire-engine boiler, and its furnace or fire-place, (which I have chosen for an example of the application of this new method to the heating and evaporating of water): AA is the boiler, which may be made of any form suitable to its use; BB is a flue surrounding the boiler, as usual; C is the up-take or passage from the space under the boiler to the flues; DD is a tunnel or flue for the flame to come from the fire-place to the boiler; EE is a place to contain the ashes; and F is a door to take them out at, (which must be kept continually shut during the time of working); GH is the fire-place: the fresh fuel is put in at G, and gradually comes down as the fuel below consumes. The part at H is very hot, being filled with cokes or coals which have ceased to smoke. J is an opening or openings to admit fresh air and regulate the fire: K is a door into the space under the boiler, which, being opened, admits air and stops the draught of the chimney, when the operation is wanted to cease. Figure 2nd is a section of the same fire-place in the other direction, in which MM is the back of the fire-place; L, the brick arch on which the fuel lies, and EE the ash-hole. Figure 3rd is an outside view of the same fire-place, showing the air-holes JJ, and the ash-hole door F; and figure 4th is a plan of the same, with part of the boiler seating, taken in the line ZZ of figure 1st. The dotted lines represent the flues, and the darts point out the direction of the flame. The fire is first kindled upon the brick arch LL, and, when well lighted, more fuel is gradually added, until it is filled up to G, and care is taken to leave proper interstices, for the air to pass either among the fuel, or between the fuel and the front wall N; and as much air is admitted at JJ as can be done without causing the smoke to ascend perpendicularly from G, which it will do if too much air is admitted at JJ. The dimensions of this fire-place are shown by the scale, and are properly adjusted for burning about eighty-four pounds weight of coals in an hour: where greater or less quantities are required to be burnt, the furnaces must be enlarged or diminished accordingly; or, if much greater, more furnaces than one must be employed. Figure 5th represents this new fire-place as applied to a furnace for melting iron and other metals, and constructed without the tunnel or perpendicular flue DD in figure 1st. (N.B. The same letters refer to the same parts in all the preceding figures.) I also construct these new fire-places so that the part GH lies sloping or horizontal, and otherwise vary the figure or form and proportions of the same, – but in all cases the principle is the same; the fresh or raw fuel being placed next the external air, and so that the smoke or flame passes over or through the coked or charred part of the fuel. I also occasionally cover the opening G, and cause the air to enter only or principally at JJ.

SECONDLY. In some cases I place the fresh fuel on a grate as usual, as at AA, figure 6th and beyond that grate, at or near the place where the flame passes into the flues or chimneys, I place another smaller grate B, on which I maintain a fire of charcoal, coke, or coals which have been previously burnt until they have ceased to smoke; which, by giving intense heat and admitting some fresh air, consumes the smoke of the first fire.

LASTLY, BE IT REMEMBERED, that my said new invention consists only in the method of consuming the smoke and increasing the heat, by causing the smoke and flame of the fresh fuel to pass through very hot tunnels or pipes, or among, through, or near fuel which is intensely hot, and which has ceased to smoke, and by mixing it with fresh air when in these circumstances; and in the form and nature of the fire-places herein mentioned, described, and delineated: the boilers and other parts of the furnaces being such as are in common use.

AND BE IT ALSO REMEMBERED, that these new invented fire-places are applicable to furnaces for almost every use or purpose.

IN WITNESS whereof, I have hereunto set my hand and seal this eighth day of July, in the year of our Lord one thousand seven hundred and eighty-five.

JAMES WATT.

Signed, sealed, and delivered (being first duly stamped), and the words 'which are 'applicable to steam engines and other purposes,' in the fifth line, – the words 'together with a current of fresh air,' in the tenth line, – the words 'and by being mixed with the 'current of fresh or unburnt air,' in the twelfth line, – the words 'air is,' in the twenty-fifth line, – the word 'admitting,' in the thirty-second line, – and the words 'and by mixing it with fresh air when in these circumstances,' in the thirty-fourth line, – being all previously interlined in presence of

JOHN SOUTHERN.
CHARLES PORDEN.

INROLLED the ninth day of July, in the year of our Lord one thousand seven hundred and eighty-five.

END OF THE LETTERS PATENT AND OF THE SPECIFICATIONS OF PATENTS.

31. Dr. Thomas Percival to James Watt,
16 September 1786, Doldowlod.

The following letter, from Dr. Thomas Percival, of Manchester, illustrates how early the problem of atmospheric pollution began to trouble the growing industrial towns, and how Watt's patent smokeless fireplace seemed to offer a solution.

'Sir

'In a conversation which I enjoyed with Dr. Priestley, who lately paid me a short but very friendly visit, I learned with much satisfaction, that you have accomplished a method of destroying smoke, which issues from fire engines, furnaces, and other works. I am solicitous to receive further information concerning a discovery, which promises to be of great importance to the inhabitants of Manchester, who appear to be peculiarly incident to pulmonic affections; and I am apprehensive will become more and more so, from the rapid increase of the cotton manufactory. The fumes which arise from the burning of velvets, are extremely acrimonious and offensive to the lungs. And they are so copious, even from a single chimney, as to scatter a shower of soot over a very considerable space.

'It is my intention to make a representation to our Magistrates, at the ensuing Quarter Sessions, of the expediency and necessity of adopting some measures for the purification of the air of Manchester. For they are guardians of the health as well as of the morals of their fellow citizens. And though works which are necessary for the prosecution of trade, ought not to be deemed nuisances, the persons who are engaged in them, should be induced, or enjoined, to conduct them in a manner as little injurious as possible, to the public.

'If business or curiosity call you to Manchester, it will afford me sincere pleasure to assure you, in person, of the respect and esteem with which I have the honour to be,

Sir,

Your most obedient humble Servant

THOMAS PERCIVAL

'P.S. I beg you will offer my best compliments to Mr. Bolton. An early answer will much oblige me.'

32. Boulton and Watt's, Directions for erecting and working the newly-invented steam engines. 1779.

GENERAL DIRECTIONS
For Building the Engine House

I. Having fixed upon the proper situation of the pump in the pit, from its centre measure out the distance to the centre of the cylinder, that is the length of the working beam, or great lever, and the half breadth of each of the great chains, as shewn by the drawing. Then from the centre of the cylinder set off all the other dimensions of the house, including the thickness of the walls, and dig out the whole ground included (to the depth of the bottom of the cellar) so that the bottom of the cylinder may stand on a level with the natural ground of the place, or lower, if convenient, for the less height the house has above ground, so much the firmer will it be. The foundations of the walls must be laid at least two feet lower than the bottom of the cellar, unless the foundation be firm rock, and care must be taken to leave a small open drain into the pit quite through the lowest part of the foundation of the lever wall, to let off any water that may accidentally be spilt in the engine house, or may naturally come into the cellar. If the foundation at that depth does not prove good, you must either go down to a better, if in your reach, or make it good by a platform of wood or piles, or both.

II. The foundation of the lever wall must be carried down lower than the bottom of the space left under the condenser cistern, (to get at the screws which fix the condenser) and two short walls must be built to carry the beams under the condenser cistern. Two other slight walls should be built one on each side, at a little distance from that cistern, to keep the earth from it, which would otherwise cause it to rot.

III. Within the house, low walls must be firmly built to carry the lower cylinder beams, so as to leave sufficient room to come at the holding-down screws, as shewn in the drawing, and the ends of these beams must also be lodged in the wall, but the platform is not to be built on them until the house is otherwise finished.

IV. The lever wall must be built in the firmest manner, and run solid course by course with thin lime mortar, and care must be taken that the lime has not been long slacked. – If the house be built of stone, let the stones be long and large, and let many headers be laid through the wall; it should also be a rule, that every stone be laid on the broadest bed it has, and never set on its edge. – A course or two above the lintel of the door which leads to the condenser, build in the wall two parallel flat thin bars of iron equally distant from each other, and from the outside and inside of the wall, and reaching the whole breadth of the lever walls. About a foot higher in the wall, lay at every four feet of the breadth of the front, other bars of the same kind at right angles to the former course, and reaching quite through the thickness of the wall, and at each front corner lay a long bar, in the middle of the side walls, and reaching quite through the front wall. If these bars are 10 or 12 feet long it will be sufficient. – When the house is built up nearly to the bottom of the opening under the great beam, another double course of bars are to be built in, as has been directed.

V. At the level of the upper cylinder beams, holes must be left in the walls for their ends, with room to move them laterally, so that the cylinder may be got in, and smaller holes must be left quite through the walls, for the introduction of iron bars; which being firmly fastened to the cylinder beams at one end, and screwed at the other or outer end, will serve by their going through both the front and back walls, to bind the house more firmly together.

VI. The spring beams, or iron bars fastened to them, must reach quite through this back wall, and be keyed or screwed up tight, and they must be firmly fastened to the lever wall on each side, either by iron bars, firm pieces of wood, or long strong stones reaching far back into the wall; they must also be bedded solidly, and the sides of the opening built in the firmest manner with wood or stone. The spring beams must always be laid 8 inches on each side distant from the working beam, to give room for the side arches.

VII. The house being finished, a wooden platform, of $2\frac{1}{2}$ inch plank, is to be laid on

127

the lower cylinder beams, and the centre of the cylinder being accurately marked on it, four holes are to be bored through the cylinder beams, for the holding down screws, and four boxes, about seven or eight inches square, and as long as the stone platform is to be deep, are to be placed perpendicularly over them. Then the stone or brick platform is to be built up to the level of the cylinder's bottom, as shewn in the drawing; it must be composed of the heaviest materials which can readily and cheaply be procured. A very solid pillar of stone or brickwork, laid in the best lime mortar, must be carried up directly under the cylinder, and must be, at least, of the diameter of the outside of the flanches; the rest of the platform may be filled up with the heavy materials, bedded solidly in a mortar of clay and sand, and well beat into their places, so as never to settle or yield.

VIII. The lever, or great working beam, is best when composed of one single log of seasoned oak; where that cannot be obtained two may be used, or four, or more; but the fewer logs it is composed of, so much the more durable will the lever be, or of so much smaller scantling may it be made. This beam is to be fashioned and mounted as in the drawing; the diagonal stays which are fastened to the arches and to the lower log, or lower edge of the beam, are to prevent the logs from sliding on one another, by the difference of the direction in which the chains act upon them when the end of the lever is up or down. These stays are to be let into the side of the beam, that the other diagonal braces may pass over them. – The diagonal braces which reach from the top of the king post to the lower edge of the beam, are intended to prevent the logs from bending or sliding on one another; they are fastened to the beam at their lower end by means of a strong square bar of iron, screwed at both ends, which passes through the beam, and serves to bind it together laterally; and they must no where else have any fixture to the beam; their screw at the top of the king post must be tightened from time to time, as required.

The gudgeon is to be placed on the top of the beam, and is not to be at all let into it, only the corners of the log may be taken off to fit the saddle-plate, and to prevent the saddle-plate from sliding on the beam; one or two pieces of hard wood, about five inches broad and a foot long, by three inches thick, may be let into the upper side of the beam, one inch deep, with their ends butting up against the saddle-plate. They must be spiked down in their places, and both them and the saddle-plate must be laid in a bed of tar and tallow mixed and used boiling hot, which will prevent the wood from rotting under them. – A clamp of oak, four inches thick, and from four to six feet long, must be spiked on the lower side of the beam; this clamp must be rounded on the edges, as shewn in the drawing; its use is to prevent the beam straps from hurting or weakening the beam in that critical place. – These beam straps must not be made out of thick bars or lumps of iron, but must be made up of a number of thin or small bars welded together; and they and all the other iron-work of the beam ought to be made of iron of the best quality; all the big pieces should be made up of smaller or thin bars in the way I have mentioned. – Upon no account whatsoever let any holes be bored through the beam near the gudgeon, nor any thing else be done which may weaken it there.

IX. The arches for the plug-tree and condenser pumps should be screwed to the beam by screw bolts, which should pass through the joints of the logs of the beam, if it be composed of more logs than one, and one bolt may generally pass above the beam; these bolts also serve to keep the beam together laterally; these arches should be made with a shoulder of two inches projection to rest on the upper-side of the beam. – The tails of the martingales of the plug-tree, and the condenser pumps, must also be secured by bolts passing through the beam in some joint, if it can conveniently be done. – The lower end of the king post should have a hollow in it to fit the gudgeon, but care should be taken that it rest upon the gudgeon, and not upon the saddle-plate. – It should be contrived that the tails of two of the great martingales should rest on the middle of each of the two logs which compose the thickness of the beam; that is, when the beam consists of four or more logs. The martingale screws should be strong, and should go down through the beam, as it is them that principally keep the beam together in the direction of its depth; near these screws must be placed the keys, or pieces of hard dry wood, which being half let into each log of the beam, prevent the logs from sliding one upon the other; these keys should never be above two inches thick, that is, one inch let into each log; they may be made in three pieces, the two outside pieces dove-tail ways, and the middle one tapering, by driving up which they are made to jam themselves in their mortoise. Or they may be made of one piece six inches broad at one end, and five at the other, so that by driving the whole in, it may check the sliding of the logs; if there are more sets of logs than two in the depth, the keys must be placed alternately on different sides of the martingale screws; care must be taken in placing the chains for the plug-tree and condenser pumps, that all the heads of the chain bolts be next the beam, and that they be far enough off not to rub on the diagonal stays, or any other thing.

X. The great chains must be made, according to the drawing, of the very best iron, and the martingales must be placed so that the adjusting screws may lye parallel to the arches, and the upper surface of the head of the martingale be at right angles to them. The holes in the martingales should be quite easy for the adjusting screws, and a washer, thinned about the outer edges, should be put under the nuts. There should be a sufficient length of chain to reach one link lower than the under end of the arch of the beam.

XI. The cap and cross bar for the piston rod should be made exactly according to the drawing, firm work of good iron; the mortoise in the cap should be made exactly to suit the mortoise in the piston rod, and the cutter or fore lock to fit them both exactly; and this cutter, above all things, should be the very best of iron, as the whole depends on it; there is always a sufficient size given it in the drawing, so that if it should fail it must be the fault of the iron or workmanship. This cutter must be kept in its place by two cross cutters, and these again by a thong of leather past through some holes in them.

XII. It will seldom happen that the plug-tree can be hung directly under its arch; you are to place them exactly in the places fixed by the drawings; that of the arch will always be found right in the general section; but the place of the plug-tree and guide posts must be taken by measuring from the nozzle, in the drawing of the working gear; a strong iron bracket with a stay must be fastened to the top of the plug-tree, in such a direction that the point of the bracket may come directly under the arch. There must be a hole in the point of the bracket, to receive the end of an iron rod reaching down from the chain, and the end of this rod must be screwed for five or six inches, and have a nut on the lower side of the bracket to adjust the height of the plug-tree by.

XIII. There should be placed upon the spring beams over the cylinder, two uprights, connected at top by a strong cape piece. These uprights serve to support a windlass with a wheel and pinion, by means of which, and a pair of tackle pully blocks hung to the cape piece, it will be easy to lift and put the cylinder, &c. in their places, and after the engine is completed, it makes it easy for the engine-men to raise the cylinder lid to pack the piston without other assistance. – The barrel of the windlass may be of oak about 6 inches diameter, and must have a square gudgeon of iron drove quite through it, on one end of which the toothed wheel is fixed; the gudgeon may be from $1\frac{1}{2}$ square to $2\frac{1}{2}$ inch square, according to the size of the engine, and the wheel about 2 feet diameter, driven by a pinion of 5 inches diameter; but these may be larger or lesser according to the weights to be commonly raised. – It is necessary to mention to those who are disposed to look on a wheel and pinion windlass as a superfluous expense, that there is no trusting to windlasses wrought by bars, and that many bad accidents have happened through the use of them, which obliges us absolutely to condemn them for this purpose.

XIV. The springs to receive and in some degree save the blow when the engine comes down too suddenly, are best made of a piece of square dry elastic timber, reaching from the plummer blocks to nine or ten inches beyond the catch pins, their size must be suited to that of the engine from six inches square to twelve or fourteen. The end next the catch pin must be sloped off on the under side for four or six feet in length, according to the size of the engine, so that their points may be one inch distant from the spring beams to which they must be bolted down by a screw bolt at the end of the sloped part, and another at the end next the plummer blocks. – The part of these springs which are struck by the catch pins, should be covered by a plate of iron, and that again by a piece of strong leather, to prevent the clattering noise they might otherwise make.

XV. The utmost attention to dimensions ought to be observed in constructing the masonry of the building, particularly in regard to heights, mistakes in them are of the worst consequences.

XVI. The condenser cistern ought to be made of the best Dantzick three inch deal plank if they can be got. If they are not to be readily got, any other good red deal or oak may be employed, but whatever kind of wood you use, be sure to cut off all the sap wood, otherwise the cistern will soon become useless. The best way of putting the cistern together, is by means of long screw bolts of iron, about $\frac{3}{4}$ square, put through the planks edgeways from top to bottom of the cistern; these screws may be 18 inches distant from one another. – The bottom may be put together in the same manner with screws, and then fixed down upon the beam or beams represented in the drawing, and supported by so many more smaller beams as may be necessary. If the cistern is not more than seven feet long, no uprights on the outside are necessary, only one about six inches square in each corner in the inside, and in no cistern ought there to be any uprights on the side next the wall. The joints of the planks should be plain joints, and put together on a strip

130

of coarse flannel soaked with a mixture of tar and tallow equal parts, used warm, or upon bullrushes. A large cock or a brass valve should be fixed in the bottom of the cistern to let off the water occasionally, and a notch about four inches deep, and 18 inches broad, with a trough fitted to it should be made in the upper edge of the cistern, to convey away the waste water. – If surface water cannot be found to supply the injection, a small pump should be fixed to bring up water from the main pump head into this cistern. – In case the water from the pit is good, and is raised to the surface, the main pump may deliver it directly into the cistern, but if the water be subject to be muddy, or mixed with sand, &c. it will be best to put it into another cistern to deposit some of that matter first. – If the pit water be vitriolic or encrusting water, it becomes necessary to use every means to procure better water, otherwise it will destroy the condenser, &c.

XVII. In making the boiler you should use rivets between 5-8ths and 3-4ths inch diameter. In the bottom and sides the heads of the rivets should be large and placed next the fire, or on the outside, and in the boiler top the heads should be on the inside. The Rivets should be placed at two inches distant from the centre of one rivet to the centre of the other, and their centres should be about one inch distant from the edge of the plate. – The edges of the plates should be evenly cut to a line, both outside and inside. It is impossible to make a boiler top truly tight which is done otherwise. After the boiler is all put together, the edges of the plates should be thickened up, and made close by a blunt chissel about $\frac{1}{4}$ inch thick in the edge impelled by a hammer of three or more pounds weight, one man holding and moving the chissel gradually, while another strikes. All the joints above water should be wetted with a solution of sal ammoniac in water, or rather in urine, which by rusting them will help to make them steam tight. After the boiler is set, it may be dryed by a small fire under it, and every joint and rivet above water painted over with thin putty, made with whiting and linseed oil, applied with a brush. – A gentle fire must be continued until the putty becomes quite hard so as scarcely to be capable of being scratched off by the thumb nail, but care must be taken not to burn the putty, nor to leave off fire until it become dry.

XVIII. In building the brick work of the boiler-setting, no lime must be used where the fire or flame comes, but a mortar made of loam or sand and clay; but lime mortar should be used towards the outside. Pieces of old cart tire or other such like pieces of iron, may be laid under the chime of the boiler, between it and the bricks, which will prevent its being so soon burnt out there. The brick work which covers the boiler top, should be laid in the best lime, which will not hurt it there but will preserve it; the mortar should be used thin, and the boiler top well plaistered with it, which will conduce greatly to tightness, if done some time before the engine be set to work. – If your lime be not of that species which stands water, it will be well to mix some Dutch or Italian terrass, or pan scratch from the salt works with it, but in any case the lime should be newly slacked. In carrying up the brick work round the flues, long pieces of rolled iron should be built in two or three courses to prevent the brick work from splitting. – Four holes at convenient places should be made into the flues, large enough to admit a boy to go in to clean them. – One of them may be over the fire door, and another right behind the damper in the back side of the chimney. This last may be as high as the flues themselves are. These holes when not in use are to be built up with nine inch brick work, and made perfectly air tight. Immediately above the brick work of the boiler-setting, a hole must be left in the chimney on the side next the boiler. This hole must be as wide as the chimney, and one foot or 18 inches high, and must have a sliding door fitted to it, to open it more or less at pleasure; the use of it is to moderate the draught of the chimney, and to prevent the flame being drawn up it before it has acted sufficiently on the boiler. A groove must be left in the brick work for the damper to move up and down in easily, which should fit flat to the face of it. – The damper may be made to move easily up and down by means of a beam or a wheel, with a counterpoise equal to the weight of the damper. The best form of a fire door is two feet long and one foot high, inside measure, to have two leaves made of boiler plates hinged on the two sides, and over lapping one another about an inch in the middle. The scantling of the frame may be three inches broad by two inches thick.

XIX. The gauge pipes may be fixed into the boiler top in some convenient place; the lower end of the longest should reach within 6 inches of the top of the flues, and the shortest should be 4 inches above it. – The feed pipe should reach two feet under the surface of the water in the boiler, and should have a valve at its lower end, to prevent the water being ever forced up through it by the steam. – Its upper end should rise seven feet higher than the surface of the water in the boiler. It should be supplied with water by a pipe from the top of the hotwater pump regulated by a cock near the feed pipe.

XX. If you have not land water that will naturally run into the condenser cistern, you must make a pool somewhere in the neighbourhood to receive the water from the hot-water pump, and reserve it for supplying the boiler and condenser cistern when the engine stands still on any occasion. – This pool may be at least 40 feet long and 20 feet wide to hold 3 feet deep of water, and pipes or troughs must be laid from its bottom to the boiler feed pipe and to the cistern. That at the feedpipe must have a cock on purpose. – It is only meant that this pool be simply dug in the earth and lined with turf, puddled, or otherwise made water tight. – If no ground within a reasonable distance be high enough for the water to run from the bottom of the pool into the boiler, than a pool may be made on lower ground and a hand pump fixed up to supply the boiler and cistern; but this ought to be avoided if possible.

DIRECTIONS
For putting the ENGINE together.

XXI. Having put the working beam together, and fastened the gudgeon to it, rest it on the plummer blocks; but do not fasten these blocks until the Cylinder is fixed.

XXII. Level the top of the stone platform, and lay the outer bottom of the cylinder down in its place, truly level, and corresponding to the holding down screw boxes.

XXIII. Apply the inner bottom upon the outer one, and set its upper joint level, by wedging betwixt it and the outer bottom if it requires it; – then cut out segments of pasteboard, such as is used for the boards of books, (not such as is composed of paper pasted together,) let these segments be of such thicknesses as the different parts of the joint may require (if it be more open in some places than in others.) Soak these pasteboard segments in warm water until they become quite soft, then lay them upon boards to dry, and when quite dry put them into a flat pan with a quantity of drying LINTSEED oil; warm the oil until the pasteboard ceases to emit bubbles of air, but take care not to heat the oil much hotter than boiling water, otherwise it will harden or burn the pasteboard. Anoint the segments on both sides with thin putty made with fine whiting and some of the lintseed oil; let the whiting be very dry, otherwise it will be difficult to mix with the oil, and N.B. that white lead will not answer in place of it.

You must as much as you can avoid using more than one thickness of pasteboard, and the segments should be a little broader than the flanch, with all the holes cut out by a chissel, but not quite so large as the holes in the iron. The segments should also be thinned at the ends where they overlap each other, so that they may form a circle of pasteboard of an uniform thickness.

XXIV. Lift up the inner bottom, and lay your segments regularly round upon the flanch of the outer bottom, then place the inner bottom upon them taking care at the same time to put a proper thickness of pasteboard in the joint under the pipe which proceeds from the inner bottom.

In like manner prepare pasteboards for the joint between the inner bottom and the cylinder, and proceed as has been directed for the other joint.

XXV. Having the cylinder ready suspended, lower it down in its place, so that the square pipe at its upper end be exactly over the pipe of the inner bottom; then with a square taper piece of iron, of a proper size, thrust into each hole, enlarge the holes in the pasteboard, so as to admit the screws. – Put in the screws and screw up the joint gradually all round, and do not screw up one side faster than another, otherwise you will be apt to crack the flanch of the cylinder or bottom: or else will make a bad joint. No screws are to be put through the cylinder flanch over the pipe, therefore that part of the joint ought to be made with the utmost care, and the pasteboard ought to be a trifle thicker there. The general thickness of the pasteboard for these joints, ought to be 3-16ths of an inch.

XXVI. Put in the holding down screws, which ought to have screws and nuts at both ends, then set the cylinder truly upright, which is done by putting a piece of wood across at bottom, and another at top, and marking upon both the centre of the cylinder at their respective places; then hang a plummet from the upper centre, and examine if the LINE be in the centre below; if it be not, you must wedge under the outer bottom, until you bring the line to hang truly in the axis of the cylinder. – The holding down screws should be screwed tight, so as to keep the cylinder in its true position, after which the screws of the joint must be again screwed up, then taken out one by one, and lapped round with a rope yarn and some putty, both under the head of the screw and under the nut, so that each screw may be air tight of itself.

XXVII. Carefully scrape, or rather scour, the rust from the sides and bottom of the cylinder; clean it well out and grease the sides with tallow. – Hang on the chains and piston-rod cap in their places – put the piston-rod into the cylinder; – suspend the piston by two half links fastened to one of the Crosses, and lower it down upon the piston rod; but previous to this, the rod should be tried into the piston, and if the hollow and convex cones do not fit one another, they must be made to do so, by chisselling and filing the cone of the rod. A lead ring an inch square exactly fitting the inside circumference of the cylinder, should be laid upon the small rim of the inner bottom, to save it in case of dropping the piston at any time; and an iron gland an inch thick should be screwed across the base of the cone of the piston rod, by means of two screws coming through the bottom of the piston and screwed into the gland. The points of these screws should be cut off so that they may not strike the bottom when the piston strikes the ring.

XXVIII. The piston being lowered down upon the rod, the lid of the cylinder is to be laid on, without the stuffing box. The end of the working beam is to be lowered down, and the piston rod cap put on the rod and forelocked fast. – The beam is to be raised, and the lid also, and an examination made whether the piston have dropped truly down to its place upon the rod. If so, the lid or cover is to be let down, and by lowering and raising the beam and piston, you will perceive whether the rod always moves up and down truly in the axis of the cylinder. It must be made to do so by shifting the plummer blocks out or in, or by shifting the martingales to one side or to the other.

The utmost care should be taken that the plummer blocks be placed both of one height; and after the beam has been some days in place, it should be examined if the gudgeon lie truly horizontal, as otherwise it will cause a most disagreeable motion in the piston rod.

XXIX. Caulk the joint round the pipe of the inner bottom between it and the pipe of the outer bottom, with rope yarn or oakum as hard drove in as possible. Screw the nozzle to the pipe of the inner bottom, making the joint as has been directed, and with the utmost care, so that the nozzle shall hang a quarter or half an inch lower at the point than at the joint, that any water condensed in it may run to the exhaustion pipe. – Put a strong wooden prop from the ground to the lower side of the nozzle, right under the perpendicular steam pipe. Care should be taken that the inside of the bottom of the nozzle, be even with, or rather lower than the inside of the bottom of the pipe which comes from the inner bottom of the cylinder, so that no water may lodge.

XXX. Put on the steam case, screwing the pannels together with a few screws. If found to be too short, it may be lengthened by means of a lead flanch put in the middle joint with a thickness of paste-board on each side of it, but if found too narrow and the deficiency upon being divided equally among all the joints amounts to more than $\frac{1}{4}$ inch to each joint, at the inner side, then a bar of iron is to be prepared of such breadth as will make up the whole deficience, and as thick as it can be put between the screw holes, in the perpendicular flanches of the steam case, and the rings on the cylinder; this bar is to be put into a joint of the steam case on the backside of the cylinder, and made tight by caulking or by pasteboard. Remember to put the middle of a pannell, opposite to the perpendicular steam pipe.

XXXI. When you have found that the steam case is of a proper diameter and length, or have adjusted it as has been directed, it is to be made tight. Make the joint between the pannels behind the perpendicular pipe, and the upper and under rings of the cylinder, by applying a proper thickness of pasteboard and putty or soft ropeing, upon the cylinder rings before you put up these pannels; or, if you perceive that the joint will admit of it, you may wind a soft rope, slackly twisted, once or twice all round the cylinder rings; then screw the perpendicular joints of the pannels together, (putting in all the screws) until the insides of the joints are quite close, or as close as they can admit of; afterwards take oakum, mixed with some putty, made with thick linseed oil; or a soft rope covered with putty, and with a caulking chissel drive it forcibly into the joint, and continue caulking in a little at a time, until you have filled the joint quite to the outside of the flanches. Remember to put oakum or soft rope yarn under the head and nut of each of the screws, as you put it in, and don't force the screws too much, lest you break the flanches; rather trust to the caulking. In like manner you are to make tight by caulking, the joints between the steam case and the upper and under rings, using a crooked chissel for the more conveniently getting at the under one.

XXXII. Put on the upper part of the lower nozzle and make its joint. Set on the perpendicular steam pipe, and ty the upper nozzle to its place; if the pipe prove too short, lead flanches, of a proper thickness, must be introduced equally above and below, to make up the length; but where-ever lead flanches are used, where hot steam comes at them, it is necessary to put a thickness of pasteboard, with putty on each side of them, and the lead should be free from tin. These lead flanches should be a little larger all round than the iron flanches, that their edges may be riveted up afterwards, when any leaks are perceived. If the pipe proves a little too long, the upper or top nozzle may be raised a little higher than its natural joint, provided the over length does not exceed an inch. The round flanch of the perpendicular steam pipe goes uppermost. Four round holes are to be drilled in the top of the upper part of the lower nozzle, corresponding to four holes in the flanch of the perpendicular pipe, and they are to be screwed together by screws, with heads within the nozzle. Five screws may, in like manner, be put in the flanch above.

XXXIII. The cross pipe is to be put on, and its joint made. The boiler steam pipe is to be screwed to one end of it, and the other shut by a plate. If any of the joints are not of a proper angle, fill them up with lead.

XXXIV. The steam is to be communicated from some convenient place of the cross pipe to the steam case, by means of a copper pipe with thin copper flanches, fixed to the cross pipe and by the steam case, by small pierced glands, with a square hole in each end, to admit the square necks of two screws, which being screwed at both end, one end must be screwed into the cast iron, first tapped for that purpose, and the other with a nut serves to keep on the gland. Another similar but smaller pipe must be fixed to the very lowest part of the steam case, bent over the flanches, and inserted into the perpendicular part of the outer bottom to fill it also with steam. In some convenient part of the outer bottom, as low down as may be, is to be fixed a waste pipe, to let out the condensed water. This waste pipe must reach down about five or six feet, and be bent upwards a little at the lower

K

end, and shut by a valve loaded with a proper weight, which will open whenever the elasticity of the steam and the weight of the pillar of water in the pipe are able to overcome the weight which shuts the valve.

XXXV. The condenser is now to be put in its place in its cistern according to the drawing. Its joints may be put together with pasteboard soaked in oil, as directed, and putty, firmly screwed up and caulked afterwards: or, any where under water plates of lead may be used, about $\frac{1}{4}$ inch thick, well fitted to the joints and puttied on both sides; after these joints made with lead, are well screwed up, and the condenser warm by fire or steam, the edges of the lead which had been left projecting a little, must be rivetted up inside and outside. A soft rope about half an inch diameter, coiled round and round until it covers the flanch, and well puttied, may be used, in default of pasteboard or lead; but either of the two former are preferable, and in every case caulking or rivetting should be used.

XXXVI. If the clack of the hot-water pump has two valves, and is not sent ready fitted, the beating or fixed part must be chiseled and filed truly flat. The pivots or axis of the valves must be from $\frac{3}{4}$ to an inch diameter, according to the size of the engine, the flat part of the iron of the valve about $\frac{1}{4}$ inch thick. – The copper facing 1-6th inch, and the iron plate under it also 1-6th inch. After the two iron plates and the copper facing are firmly rivetted together, they are to be heated red hot, laid on their place, a short piece of end wood set above them, and beat down by some blows of a sledge hammer. – The pieces of iron the pivots move in, are to be fixed by means of pins of iron half inch or $\frac{3}{4}$ square, screwed into the cast iron of the clack, passing through a square hole in the pivot pieces, and forelocked above by spring cutters. Every one of these parts ought to be made very secure and firm. – A guard to prevent these valves from over opening, is to be fixed in the hot-water pump, according to the drawing, this guard may be about an inch thick and should not touch the edges of the valves, but catch them on the flat part behind. – The cast iron face of the eduction pipe foot is also to be made flat for the valve there to beat against. – The pivots of this valve should be one inch diameter. The thickness of the iron and copper the same as for the others. – The ends of the valve should be one quarter of an inch clear of the sides, and one half inch clear of the bottom of the place it plays in. – The pivots should be sunk into the cast-iron of the sides until their lower edge be within one quarter of an inch of the opening of the beating part. – They should have one inch of hold of the iron at each end, and have no play in that direction. – In the lid or clack door for this valve there should be a groove for the axis of the valve, so that it may not touch it when the lid is screwed on. – The pivots should not be confined close against the beating part, but should have a quarter of an inch of play in that direction, as the air makes its escape partly at the hinge. – The valves of the air and hot water pump buckets are to be fitted in the same manner, remembering to make the pivots proportionable to the size of the valves.

XXXVII. The condenser being fixed in the cistern at the height below the nozzle, and distance from the centre of the gudgeon or of the cylinder, shewn by the drawings, and so that the middle between the centres of the pumps shall be directly under the middle of the working beam, and the line between these centres at right angles to the the beam, the copper eduction pipe is to be fitted to its place. It is to be screwed to the flanch of the short pipe under the nozzle by means of a loose flanch of hammered or cast iron applied on the underside of the copper flanch of the pipe. The outside diameter of the loose flanch should be the same as that of the flanch on the nozzle, and its inside diameter should be one inch more than the outside diameter of the bent copper pipe, and should have its inner angle taken off a little on the side next the copper flanch, lest it should cut that flanch, or crack the soldering. – If the loose flanch be made of hammered iron, it should be ¾ of an inch thick, and the holes should be drilled and not punched; and in the same way you are to proceed with the joint at the foot of the eduction pipe.

XXXVIII. Having carefully tinned the inside of the upper end of the wide or perpendicular part of the eduction pipe, and also the outside of the brass ring which goes within it, the ring is to be put into its place, and being heated, the joint is to be run with fluid tin solder, after which four or more holes may be drilled through both the copper and the brass, and some copper rivets put in them. – The spiggot and fosset joints are to be secured as follows: An iron ring three or four inches broad, and half an inch thick, is to be put red hot on the outside of the fosset part, so that by its contraction in cooling it may grasp it firmly. – The spiggot part is then to be put into it and made tight by caulking in soft roping and putty. The proper width of a joint for caulking is 3-16ths of an inch at the wide or open end, and drawing quite close at the inner end; but will answer although a little wider or narrower.

The joint of the bent part of the eduction part to the perpendicular part at the brass ring, is also to be done by caulking.

When the engine is set to work, if any of the spiggot and fosset joints shew a disposition to slide or move, that may be cured by putting screw hoops round both the spiggot and the fosset part near the joint, and pulling the joint together, by means of two screws connecting the screw hoops.

Care must be taken in putting the eduction pipe together, to keep the brased joint upwards, so that if any defects appear they may be cured by tin solder.

When the eduction pipe is all put together, a hole is to be cut for the short fosset pipe of the injection, as shewn in the drawing. This hole must be cut, so as to fit the outside of the fosset accurately. – The fosset pipe should point up the eduction pipe in such manner that the injection water may strike the upper side of the eduction pipe within about two feet of the nozzle; but care should be taken that it do not spout too low, otherwise it may, by the bent pipe, be reflected up against the exhaustion regulator, which will be very hurtful. – The fosset pipe being adjusted to its proper position, and the knee of the eduction pipe tinned round the hole, the fosset is to be fixed in its place, either by a strong body of plummer's solder, or by a copper bosse or case run full of the same solder heated to a dull red heat. – The upper edge of the inner end of the fosset should only go half an inch within the eduction pipe, and the nozzle of the injection not quite so far. – The injection pipe being set in its true position, the joints soldered with plummer's solder, and its valve soldered on, a hole is to be cut for the blowing pipe fosset at or about the level of the valve of the injection, but not lower, otherwise the engine will blow at the injection, and heat the cistern. – The fosset for the blowing pipe is to be fixed by soldering, or by a bosse, as directed for the injection, and its inner end ought not to go more than one or two inches within the eduction pipe, according to the diameter of that pipe. – The blowing pipe may then be put together, and its valve soldered on, taking care that the pipe be of such a length that its valve may be 6 inches under the surface of the water in the cistern. – Care must be taken that the stems of both the blowing and injection valves stand truly perpendicular when fixed in their places. – The injection and blowing pipes are to be fixed in their fossets by caulking as directed. Care must be taken that no tin from any of the solderings be left in the pipes.

XXXIX. The condenser pumps must be fixed down by means of screws passing through the bottom of the cistern, and the beam under it, as shewn in the drawings; and it must be remembered, that its disposition to rise is very powerful, and that if it has any play it will be sure to spoil the joints of the eduction pipe, or perhaps break it. The hot water pump must have a strong prop under it, and be tied down as well as the other. A beam of deal, nine or ten inches square, must be put across the cistern, near the air pump rod, to support a pair of shears or uprights, for a pump brake for that pump, to examine the tightness of the joints by. This brake must have an arch and a chain with a hook, to join it to the chain of the air pump rod, when in use. The buckets of the condenser pumps must be surrounded by a plaited rope made of rope yarn, of such breadth as to fill easily the interstice between the bucket and the pump barrel. A pudding link chain, four foot long, must be fastened to the top of the sliding-rod of the air pump, and to the other part of that rod which reaches up to the working-beam; its use is to suffer the engine to work without unloosing the hook of the pump brake, when trying experiments on the tightness of the engine.

XL. The stuffing box of the air pump must be packed with a small soft rope wrapt round the rod, and forced down into the box pretty tight, but so that the rod may move easily. A flat round piece of wood about $1\frac{1}{2}$ inch thick, fitted easy to the inside of the box, and to the outside of the rod, must be put above the stuffing and screwed down by the gland. There need be no screws put to hold down that side of the air pump lid which is over the connecting box; those on each side of the box are sufficient, if care be taken in making the joints. In like manner the lid or clack door of the lower valve of the eduction pipe foot needs only two screws, one at each end. In the bottom of the air pump must be placed a ring of hammered iron, with three or four feet, for the bucket of the pump to rest upon when at its lowest, i.e. when the lower edge of the packing of the bucket is within one inch of the under end of the working barrel; this ring must be fixed so that it may not turn round and come in the way of the lower valve of the eduction pipe. An upright, six inches square, must be fixed from the bottom of the cistern, near the injection, to screw that pipe to; and its upper end must be fastened to the beam which carries the lever, and the end of the working gear of the injection. This upright should be fixed firmly, and the injection pipe should be fastened to it, by a stirrup with screwed ends, grasping the neck of the valve, going through the upright, and having nuts behind it. Any motion in the injection pipe will be apt to loosen or crack the joints of it, therefore it must be firmly fastened.

XLI. Guards must be fixed over the injection and blowing valves, to prevent their over opening; for the knob of the spindle which stops them, by the bridge of the valves, is not to be trusted, and may therefore be cut off, which will give the convenience of taking out the fly part of the valve at pleasure. An S hook of iron is to be fitted into the eye of the valve, so as to have no motion there, and the rod which pulls it open is to have a hole in its lower end, for the upper end of the S to play easily; if it be allowed to have motion in the eye of the valve, it will soon wear it out. Guards are also to be fixed over the valves on the air pump lid, to prevent their over opening; these guards may be fixed by means of two of the screws which fasten on the lid.

XLII. The guide posts, or Y posts, of the plug frame, are to be fixed exactly according to the drawing sent for that purpose; and the cross swords which slide in the guide posts should be of oak or beech, two inches thick and eight or nine broad. The plug-tree itself should be of hard, straight grained, seasoned oak, the holes $1\frac{1}{4}$ inch diameter, bored off both sides by a centre bit; for if you bore them by an augre, they will be apt to break out into one another; care should be taken to bore a sufficient length of the plug. The opening horns, or arches of the Y shafts, which act upon the levers of the regulators, must be bent exactly to the curves of the FULL SIZE DRAWINGS sent for them. This is best done by taking a piece of soft iron, an inch broad and three sixteenths thick, and bending it cold until its hollow side exactly fit the drawing, and by applying this mould to the arch, while red hot, you can set it truly into form. These moulds should be carefully laid up, lest, by any accident, the arches should require repairs. To fix the Y shafts, make the levers of both regulator spindles truly horrizontal, and so long as just to reach to their proper places on the Y shafts, and then the lower side of the exhaustion lever, and the upper side of the steam lever will point to the axis or centres of their respective Y shafts. The coupling brasses for the Y shaft pivots or gudgeons, are to be fixed one inch from the inside of the guide posts, and the centres of the pivots are to lye exactly in the line of the inner side of the rabbits or grooves, in which the swords move, (as drawn). A piece of wood, with a slit in it three inches wide and about three feet long, having holes in it, like an old-fashioned plug-tree, must be placed to receive the opening horn and lever of the steam regulator, and by means of wooden pegs, one inch diameter, put through its holes, and saddles of leather laid above them, regulate the opening of the steam regulator. To prevent shaking and noise, the lower end of this piece of wood should rest on the ground, in the floor of the cellar. The lower end of the guide posts must be fixed upon sills parallel to the working beam; otherwise the weight of the exhaustion would fall upon them and shake them every stroke. The floor over the eduction pipe should be easily moveable, that the pipe may be come at. There should be a window in the door which leads to the condenser, to give light to the plug-frame. The weight which hangs to the detent of the exhaustion, and serves to raise the arch and open that regulator should be of lead cast on the rod, and square pieces of lead, with a notch in them, to admit the rod, may be laid on, if the weight proves too light. Some oakum should be laid between these saddles to prevent noise; a box, eighteen inches square, and two feet deep, should be fixed about the blowing pipe, to prevent the hot water from mixing with the cold in the cistern, but there should be a few holes in the bottom of this box to suffer the water to go out below; this box should rise six inches above water.

XLIII. Care should be taken that both the regulators fall into their seat without touching sooner on one side than the other; and if the copper cones, under the regulators, be not already rivetted or screwed to them, it should be done before you begin, but avoid bending the valves in so doing; some threads of oakum, well puttied, should be lapped round the necks of the regulator spindles, beyound the shoulders, to keep them steam and air tight; but this should be done in such manner as not to prevent the spindles from going quite home to their shoulders, otherwise the regulators cannot fall right in their places.

XLIV. The brass of the cylinder stuffing-box should be fixed in its place, and the upper or thin edge of it set out against the sides of the iron part. When the piston rod plays truly up and down in the axis of the cylinder, put on the stuffing-box, and screw it down by its flanch; then pack the box with soft rope yarn, wrapt round the rod, until you have nearly filled the box, then take a collar of deal wood, two inches thick, made easy for the rod and for the box; divide it in two by its diameter, lay it on the top of the stuffing, and apply the gland above it; as you go on with the packing, melt some grease and pour amongst it, and when finished screw down the gland moderately tight.

XLV. The cylinder lid is to have no screw holes over the square pipe; its joint is to be made with pasteboard, puttied on the lower side but not on the upper side, and the lid being greased with tallow, the pasteboard will not stick to it, but will lye in its place when the lid is raised. Two long iron rods, with hooks at their lower end, should be hung to eye bolts in the spring beams, so that when the lid is raised about three feet from the cylinder, these hooks may be put into two opposite screw holes, to support the lid at that height while you pack the piston.

XLVI. To pack the piston, take sixty commonsized WHITE or untarred rope yarns, and with them plait a gasket or flat rope, as close and firm as possible, tapering for 18 inches at each end, and long enough to go round the piston, and overlap for that length; coil this rope the thin way as hard as you can, lay it on an iron plate, and beat it with a sledge hammer until its breadth answers its place; put it in and beat it down with a wooden driver and a hand-mallet; pour some melted tallow all round; then pack in a layer of white oakum, half an inch thick, then another rope, then more oakum, so that the whole packing may have the depth of about four inches, or only three inches if the engine be a small one. Cast segments of a circle of lead, about 12 inches long, three inches deep, and $1\frac{1}{4}$ inch thick, fitted to the circle of the piston, and cut down square at both ends; lay them round upon the packing as close as they can lye to one another without jamming, and screw down the piston springs upon them; the piston springs should be bent downwards at the end next the piston rod, and a little mortoise should be cut in the cast iron there, for the bent down point of each of them to lodge in, which will prevent their coming forwards to touch the cylinder. Previous to the piston being put into the cylinder, the

140

hollows among the crosses should be quite filled up with solid pieces of deal wood, put in radius fashion. The packing of the piston should be beat solid, but not too hard, otherwise it will create so great a friction as to hinder the easy going of the engine. Abundance of tallow should be allowed it, especially at first; the quantity required will be less as the cylinder grows smooth.

XLVII. The joints being all made, the regulator valves in their places, and their covers screwed on, but no water in the condenser cistern, admit steam, and when the cylinder and steam case are thoroughly warmed, screw up the nuts of all your screws, and caulk the pasteboard or oakum of such joints as may require it, with a caulking chissel, until you find that every thing about the cylinder is perfectly staunch; then pour three or four feet deep of water into the hot-water pump; stake down the injection and blowing valves, and also those on the air pump lid, and let the steam into the condenser, which will shew the defects or leaks, if there be any.

XLVIII. Screw on the steam gauge to the steam case near the nozzle, and behind the engine-man's place, pour as much mercury into it as will half fill the open leg; put a float on it, broad at bottom, but very slender in the stem; cut the float or index off close to the end of the open tube, and fix a scale to it, reckoning every half inch the float rises equal to an augmentation of the elasticity of the steam, corresponding to the supporting a column of mercury an inch high, because the surface has sunk as much in the one leg as it has risen in the other. – Solder a small copper fosset pipe, to fit the copper communicating tube of the barometer, into the eduction pipe, 12 inches under the fosset of the blowing valve, and on the opposite side of the eduction pipe; place the barometer in the door way to the condenser on the further side from the plugtree, so that the engine man may see it when at his station; join the copper tube to it, by pouring melted sealing-wax into the copper cup at top, fill the short leg of the barometer with mercury, within four or five inches of its top, and put a light float in it, long enough to reach to the top of its frame.

XLIX. Fill the condenser cistern, shut the lower regulator, and there being no steam in the cylinder or its communication with the boiler being cut off, take off the bonnet or cover of the exhaustion regulator, shut that regulator, and work the air pump by means of the brake. If then you find that air enters by the regulator, pour some water on it, and continue pumping until you have raised the barometer, i.e. sunk its float to 27 or 28 inches; leave off pumping, and observe if the vacuum continues good, or is a long time in being destroyed. If it loses fast, seek for the leaks which must be somewhere in the eduction pipe, and will make a noise if touched with a wet hand; (observe if the condenser moves by the pumping, and secure it.) After having cured these leaks, you may try the tightness of the cylinder, by staking the working beam, so that the piston cannot descend; then taking the cover off the cylinder, open the exhaustion regulator, and shut the steam regulator; on beginning to pump, you will perceive if the piston be tight, if it is not, it may be beat a little, and some water being thrown upon it, and on the steam regulator, whatever air enters, must be by leaks, which must be fought for and cured by screwing or caulking in oakum. – N.B. A critical tightness in the piston cannot be obtained until the engine has gone a few days, without beating it too hard, to permit the engine to move easily. – When you can detect no more leaks in this way, the steam must be admitted, and the same examination made as before.

L. The piston chain should be so adjusted, that it may descend within one inch of the lead ring at bottom when the springs are pressed down by the catch pins, and that, when it is at its highest its upper edge may be level with the square opening at top, so that no water may lodge there, but may run down the perpendicular pipe; and the engine should always be made to work full stroke, otherwise it will spoil the cylinder. – A collar of soft rope should be lapt round the piston rod under the lid to prevent the piston striking it if it should rise with a jump. And if the cap of the piston rod does not touch the gland of the stuffing-box when the catch pin have pressed down the springs above, a collar of iron must be fitted on the rod to make up the deficience, and to help to save the blow if the chains should give way and the piston fall; for, though it should break the cylinder lid, that is a much smaller damage than the bottom would be, as it may be clasped or otherwise mended.

LI. There ought to be cleets or strong brackets of wood firmly bolted to the dry pump rods, and beams put across the pit at proper distances to receive them in case of the accident of their breaking.

LII. After the engine has been set a-going, and has gone a few hours, the holding-down screws should be screwed tight, and so from time to time as they become slack; and in like manner all the other screws about the cylinder or nozzles should be screwed up as they slacken, and the joints caulked and puttied where they require it.

Directions for Working the Engine.

LIII. It being necessary that the uses of the several regulators be thoroughly understood by those who attend the engine, we shall begin by describing them.

In the lower nozzle or regulator box are two regulating valves. When the upper one is opened, it admits the steam, from the perpendicular steam pipe into the cylinder below the piston, and thereby permits the piston to ascend, or in the engine man's phrase, allows the engine TO GO OUT OF THE HOUSE; this regulator we call the STEAM REGULATOR. The lower regulator, which is placed in the bottom of the nozzle or regulator box, when open suffers the steam to pass from the cylinder into the air pump of the condenser, and thereby a vacuum is produced in the cylinder; this valve is called the EXHAUSTION REGULATOR. – There is a third regulating valve, called the TOP REGULATOR, placed in the cross pipe at the upper end of the perpendicular steam pipe, which serves to proportion the Quantity of steam to be admitted from the boiler, to the load of the Engine; so that when the load is less than ten pounds and a half on the inch, the steam in the upper part of the cylinder, which presses on the piston, may be less dense or weaker than the steam in the boiler, and consequently a smaller quantity may be employed to do the work than what is required when the engine is fully loaded. This regulation may be effected in two ways; either by opening the top regulator fully, at the beginning of the stroke, and shutting it before the piston arrives at the bottom; or by opening it so far as just to give the piston a sufficient velocity, and keeping it open until the end of the stroke.

LIV. The engine being supposed in motion, the operation of these valves will be as follows, when the piston is at the bottom of the cylinder, and the exhaustion regulator is shut, if the steam regulator be opened, the steam will pass through the perpendicular steam pipe and that regulator from the part of the cylinder above the piston into the part below it, and the steam thereby becoming equally strong or dense above the piston and below it, will give no resistance to the ascent of the piston which will therefore be pulled up by the superior weight at the pump end of the working beam.

When the piston is come to the upper end of the cylinder, the steam regulator must be shut, the exhaustion regulator opened fully, and at the same instant the top regulator opened so far as to admit the proper quantity of steam, (the degree of this opening must be determined by experience): The steam contained below the piston will then rush from the cylinder through the exhaustion regulator into the vacuum or empty space in the eduction pipe, where it will meet the jet or stream of injection water which will instantly condense or reduce it to water, and thereby exhaust or empty the cylinder of steam.

The steam in the upper part of the cylinder being no longer ballanced by steam below the piston will press upon it by its elasticity, and it will begin its motion downwards; as the piston moves downwards the steam in the upper part of the cylinder will become less dense than that in the boiler, which will therefore enter the upper part of the cylinder by the opening of the top regulator, and will maintain the steam in that part of the cylinder in a proper degree of density or strength to give the necessary velocity to the piston, and to press it to the bottom of the cylinder; but if the engine be underloaded it will be necessary to shut the top regulator a little before the piston is at the end of its stroke. It has been observed that the precise time at which the top regulator should be shut must be determined by experience, no certain rule can be given, because it depends upon the degree to which it is opened, and upon the load of the engine at the time; but it must always be shut sooner than the exhaustion regulator, which is kept open to the end of the stroke.

The injection valve should be opened a little before the exhaustion regulator, that the exhaustion pipe and the water remaining from the last stroke may be cold when the steam enters, by which means the condensation will be performed more suddenly; and the injection should be shut very soon after the piston begins to descend, observing however to let it play so long that the degree of vacuum shewn by the barometer may be greater in the latter part of the stroke than in the beginning of it. The opening or adjutage of the injection pipe should be proportioned to the load of the engine, so that the proper quantity of water may enter in about one second of time, and as the load encreases, the opening must be enlarged.

LV. The eduction pipe serves to convey the injection water and condensed steam to the foot of the air pump of the condenser; the injection pipe enters it at its knee, and spouts along the horizontal part of it; and from its side issues the blowing pipe, the use of which is to empty the eduction pipe of air and water when the engine is put in motion after it has been stopt at any time. At the bottom of the eduction pipe is a hinged valve

143

or clack which permits the water and air to pass into the air pump, but prevents it from returning. This valve should be very tight; it is called the valve of the eduction pipe foot.

LVI. The AIR PUMP is the lowermost and widest pump of the condenser. When the steam enters the eduction pipe it spoils the vacuum for an instant, and then presses upon the water in the lower part of the eduction pipe, and forces a part of it into the air pump; as the piston of the cylinder descends, the bucket of the air pump ascends, carries up along with it the hot water which was above it, and leaves a vacuum under it, into which the remaining injection water enters; first because it stands higher in the eduction pipe than in the air pump; and secondly because the vacuum in the eduction pipe is not quite so complete as in the air pump.

The water raised by the air pump bucket passes through the clack of the hot-water pump into the vacuum produced by the rising of the bucket of that pump, which is raised at the same time with the bucket of the air pump, and no part of it will come out at the valves on the lid or cover of the air pump, unless the bucket of the hot-water pump is not tight, or an overplus quantity of water enters the eduction pipe or condenser by leaks; for if there be a sufficient empty space left by the bucket of the hot water pump it is evident that the water will rush into it, and fill it before it can open the valves on the lid, which are kept shut by the pressure of the atmosphere so long as there is any degree of vacuum in the upper part of the air pump, or that part of the hot water pump which communicates with it. When the air pump bucket descends, it leaves a vacuum behind it, because the water is retained by the hot water pump; and the water in the lower part of the air pump passes through the valves of the bucket, which lifts it up the next stroke as before.

The hot water pump raises the water high enough to let it run into the boiler by the feed pipe, or into a reservoir to be cooled, and so to serve the purpose of injection a second time.

LVII. The barometer serves to shew the degree to which the cylinder is exhausted of air and steam; it consists of a longer and a shorter tube of iron, both of one diameter and truly bored, and joined together at bottom by a bent iron pipe; it should be fixed up perpendicular, and should be filled with mercury until it stands 18 inches deep in the shorter or open leg; a light float of wood something like a gun-rammer should be put into the short leg, and cut off even with the top of the scale when the engine is at rest, and the eduction pipe filled with air; the scale is divided into half inches, which correspond to inches on a common barometer, because for every half inch the mercury rises in the long leg, it falls half an inch in the short leg, which, added together, make one inch difference of height; a pipe from the top of the long leg is joined to the eduction pipe, below the blowing valve, for were it fixed higher, steam might come through it and loosen the cement that connects the pipe and the barometer. – When the mercury in the common barometer stands at 30 inches, it should stand at $28\frac{1}{2}$ inches in this barometer, if your engine be in order, or in proportion at other heights.

The steam gauge is a similar instrument, in which the steam presses up a column of mercury proportioned to its elasticity. When an engine is underloaded it ought to be wrought with steam able to support one inch of mercury, and when full loaded it ought

144

not to exceed two inches; but if the engine be loaded to more than ten pounds and a half on the square inch of the piston, the strength of the steam must be increased accordingly.

It is never adviseable to work with a strong steam where it can be avoided, as it increases the leakages of the boiler and joints of the steam case, and answers no good end.

LVIII. A very important article is the proper packing of the piston, directions for doing which have been already given, (XLVI) but as that part may not come into the engine-man's hands, it is proper to repeat it here: Take sixty white or untarred rope yarns, and with them plait a gasket or flat rope, as close and firm as possible, tapering for eighteen inches at each end, and long enough to go round the piston and overlap for that length; coil this rope the thin way as hard as you can, lay it on an iron plate and beat it with a sledge hammer until its breadth answers its place; put it in and beat it down with a wooden driver and a hand mallet; pour some melted tallow all round; then pack in a layer of white oakum, about half an inch thick, then another rope and more oakum, so that the whole packing may have the depth of four inches, or only three inches if the engine be a small one; soak the whole well with melted tallow, and after having beat the packing moderately, lay on the piston leads; put on the springs and screw them down. In a new engine the piston must be examined after about twelve hours going, and be beat a little and fresh greased; but you must be careful not to pack or beat it too hard; otherwise it will create so much friction as almost to stop the engine.

LIX. The buckets of the air and hot water pumps are to be packed with a flat rope, wrapt round them edge ways; and the ends of these gaskets should be made fast by being drawn through holes made in the buckets, for that purpose, and secured there by wooden pegs hard drove in. – The gaskets should be well smeared with tallow before the buckets are put in, and they should not fit the pumps too tight, as their sticking is very troublesome, especially at first.

The stuffing boxes of the cylinder and air pump are to be packed by wrapping a soft rope round the rod, and beating it in until it nearly fills the stuffing box, remembering to soak it well with tallow as you go on; above this rope lay on the wooden collar, and screw the gland down upon it moderately tight.

LX. To set the engine a going, raise the steam until the index of the steam gauge comes to three inches on the scale; when the outer cylinder is fully warmed, and steam issues freely on opening the small valve at the bottom of the syphon or waste pipe, which discharges the condensed water from the outer bottom, open all the regulators; the steam will then forcibly blow out the air or water contained in the eduction pipe, by the blowing valve, but cannot immediately take place of the air in the cylinder itself; to get quit of it, after you have blown the engine a few minutes, shut the steam regulator, the cold water of the condenser cistern will condense some of the steam contained in the eduction pipe, and its place will be supplied by some of the air from the cylinder; open the steam regulator and blow out that air; and repeat the operation until you judge the cylinder to be cleared of air; when that is the case, shut all the regulators and observe if the barometer shews that there is any vacuum in the eduction pipe; when the barometer gauge has sunk three inches, open the injection a very little, and shut it again immediately; if this produces any considerable degree of vacuum, open the exhaustion regulator a very little way, and the injection at the same time, if the engine does not commence its motion, it must be blown again and the same operation repeated until it moves; if the engine be very lightly loaded, or if there is no water in the pumps, you must be very nimble and shut the exhaustion and top regulators, so soon as it begins to move quickly, otherwise it will make its stroke with great violence, and perhaps do some mischief. To prevent which, open the top and exhaustion regulators, only a little way, and put pegs in the plug-tree, so that they may be sure to shut these regulators long before the piston comes to the bottom.

If there is much unbalanced weight on the pump end, you must also take care to put a peg in the ladder which guards the steam regulator lever, so as to allow that regulator to open only a little way, and so to lessen the passage for the steam, when it enters to fill the cylinder, otherwise the rods, &c. at the pump end may descend too fast and be pre-judicial; if you find after a few strokes, that the engine goes out too slow, the steam regulator may be opened wider. In order to regulate the opening of the exhaustion regulator, you should have pieces of board of various thicknesses, to put under the weight which pulls it open, by means of which it may be made to open more or less at pleasure, and the top regulator may be managed in the same manner.

LXI. Should the engine work with too great violence on account of its being under-loaded, you may correct it by giving the top regulator a lesser opening, and shutting it at such a part of the stroke as will just give the piston sufficient force to come to the bottom. Whenever the top regulator is used, the exhaustion regulator should be thrown fully open every stroke, in order to give a free exit to the steam, on which a great part of the good effects of the top regulator depends.

The engine should always be made to work full stroke, that is until the catch-pins come within half an inch of the springs on each end, which is easily managed by an attention to the pegs. Care must be taken, that the piston rise high enough in the cylinder when the engine is at rest, to spill over into the perpendicular steam pipe any water which may be condensed above it; for if any water remain there, or in any other part of the cylinder while it is working, it will very much encrease the consumption of steam. When the engine is to be stopt, shut the injection and secure it, put a peg in the plug-tree to prevent the exhaustion regulator from opening, and take out the peg on the other side, so as to allow the steam regulator to open and to remain open; otherwise you may have a partial vacuum in the cylinder, and it may be filled with water from the injection or leakages, which is a troublesome accident. – The top regulator should also be open while the engine stands.

When an engine is in tolerable good order it will bear to stand ten minutes, and go to work again without blowing afresh, and though it has stood two or three hours, if there has been any steam issuing from the boiler, and no air has been admitted into the cylinder, it will generally go off with once blowing for about a minute.

LXII. If you find, after following the above directions, that the engine does not go to work, shut the exhaustion regulator, and give some injection, if it then makes no vacuum, it is likely there are air leaks about the eduction pipe; if it does make a vacuum, which remains but a short time, it may be owing either to air or water leaks, these may be distinguished by blowing as before, and shutting the lower regulator for about a minute, without giving any injection. If upon opening it again, it throws out a good deal of water at the blowing pipe before it blows steam, it is certain that it either has some leak in the condenser under water, or that the injection or blowing valve does not shut close, every joint should be examined, and also the valve at the foot of the eduction pipe.

If after blowing as before, you find that immediately on opening the exhaustion regulator, a quantity of air is thrown out at the blowing valve, the leak is in the eduction pipe some where between the surface of the water in the cistern and the nozzle. The particular place of these leaks may be found, by emptying the cistern of water, putting three or four feet deep of water into the hot water pump, and staking down the blowing and injection valves with those on the air pump lid; then if steam be admitted into the eduction pipe, it will come out at the leaks and point them out. – If not found out in this way, apply the brake to the air pump, taking care first to put some water on its bucket, and then by working that pump by hand, you will probably on an attentive examination observe where air goes in, which may be known more distinctly by wetting the place suspected.

If upon shutting the lower regulator and making a vacuum in the exhaustion pipe by pumping, or by injection, you find that vacuum continues good for a considerable time, then the fault does not lie in the eduction pipe, but in the nozzle or joint of the cylinder bottom, where it must be sought for.

In these examinations by pumping it is proper to take off the bonnet or cover of the exhaustion regulator, and to examine if air enters at that regulator, if it does, and only in small quantity, throw some water on the regulator while you are examining the eduction pipe; and when the leak is suspected to be in the bottom joint of the cylinder, or in the lower nozzle, you must throw some water on the steam regulator and also on the piston, then by pumping and strict examination you will soon find where the air enters. When you are examining the tightness of the piston by pumping, you must stake the beam, so that the piston may not descend.

LXIII. If in course of working, you do not find the vacuum keep good, and the engine goes sluggishly, or stops and requires to be blowed frequently, you must examine whether an uncommon quantity of air or water issues at the hot water pump, or if any comes out at the valves on the air pump lid; if the quantity of air is great, the engine has some air leak, and if the quantity of water be great, and is rather cooler than usual, it proceeds from a water leak in the condenser; if the quantity of water be great, and at the same time very hot, it proceeds from a bad piston, or from the steam regulator not shutting close.

The engine will also go badly if the air pump or water pump buckets or clacks strip the water, that is let it pass by them; you will know if this be the case with the water pump bucket, by observing whether the water follows down after it at the return of the stroke, and leaves a part of the pump empty; if it does not, either the bucket strips the water, or the engine receives water in some way which it ought not.

LXIV. Attention ought to be given to feeding the boiler in a regular manner, that it may not be spoiled, nor steam wanted. When there is too much water in the boiler, the engine will not work regular, and if there is too little, the sides of the boiler will be burnt by the flame in the flues. – If by accident it should at any time run a little too low, the feed should be augmented, so as to fill it gradually; for if you run in too much at once, you will check the steam and stop the engine; but if it be run very low, stop the engine, open the puppet clack, and fill the boiler from the pool or reservoir if you have one; otherwise fill it by working the air pump, having first staked down the valves on its cover, and opened the injection valve. – In working the engine the steam ought to be strong enought to make the index of the steam gauge stand half an inch high at least, otherwise air will enter at the joints of the boiler, &c. and spoil the vacuum, so as to cause a good deal of trouble to get quit of it again. Therefore if you perceive the steam gauge to be lower, stop the engine until it rises again. By a little attention, you will find the proper opening of the feeding cock for any rate of working.

LXV. Let all the coals employed to feed the fire, be thoroughly watered just before they are thrown on, as that will prevent their being swept into the flues by the draught of the chimney.

The fire should be kept of an equal thickness and free from open places or holes, which are extremely prejudicial, and should be filled up as soon as they appear; if the fire grows foul and wants air by clinkers collecting on the bars, they must be got out with a poker, but the fire should be as little disturbed in that operation as possible, and the greatest care taken not to make any coals or coaks fall through, which are not thoroughly consumed; it is very common for a fourth of the whole coals to be wasted in this manner, by mere carelessness. When the fire is newly made, the damper should be raised a little, so as to let off the smoke freely, but should be let down to its proper place so soon as the smoke is gone off. The air door in the chimney should be always open more or less; it prevents the flame from being sucked up the chimney, and very considerably increases the effect of the coals. Once a month, the boiler and flues ought to be cleaned, or oftener if the water be very subject to incrust the boiler. Every morning the ashes ought to be taken out, the engine house swept clean, and a view taken of every part of the engine, to see that nothing be working out of its place, or want oiling. Particular attention ought to be paid to the bolts and cutters of the great chains and piston rod, so that none of them get loose.

LXVI. Once every week let the top of the cylinder be taken off, and also the springs and leads of the piston; let the packing be beat down moderately, with the driver and mallet, and fresh oakum, or a gasket added when necessary. For every foot the cylinder is in diameter, pour two pounds of melted tallow on the packing, before you put in the leads, and for two or three hours after you have added the tallow keep the piston from rising quite to the top of the cylinder, by laying two pieces of wood three inches thick on the outside springs, that the tallow may not be split off before it has time to soak into the packing. At the same time you pack the piston, you should examine the state of the condenser, and rectify any thing you find amiss; and while these things are doing the pitwork should not be neglected, that one stoppage may serve for all.

LXVII. The regulator valves should be examined from time to time, and a little fresh oakum should be lapt about the necks of their spindles to keep them air and steam tight. The stuffing-boxes also should be minded, and no steam suffered to escape any where; its escaping is a mark of slovenliness, and a material injury both in extra-consumption of coals, and in the destruction of the iron and wood-work.

An engine, when in good order, ought to be capable of going so slow as one stroke in ten minutes, and so fast as ten strokes in one minute; and if it does not fulfil these conditions, somewhat is amiss that can be remedied.

The hot water should issue of the heat of 96 degrees of Fahrenheit's thermometer, that is blood warm, when the engine is in excellent order, and should never exceed the heat of 110 degrees, unless when the injection or cold water is hotter than 70 degrees, and in that case the vacuum will not be good.

LXVIII. At the end of the horizontal steam pipe next the boiler is fixt the STEAM REGULATOR, the use of which is to shut off the steam while any thing is doing about the top regulator, or other parts connected with it. It may also be used to stop the communication with one boiler, while another is in use.

LXIX. At the first setting an engine to work, it frequently happens that there is a difficulty in procuring a sufficient quantity of cold water for condensation, and it also frequently happens that there is something or other amiss, which may occasion the engine to be a long time in setting to work, and by the repeated blowing, the water in the cistern gets too hot to serve for condensation. In such cases a great deal of trouble may be saved by exhausting the air from the cylinder by working the air pump by the brake, having first opened the exhaustion regulator and shut the steam one. – And in any case when the engine does not go readily to work by blowing, and the quantity of injection water is limited, it is best to set on by pumping, and even to assist the engine for a stroke or two by the same means, if it be fully loaded. – As the bucket of the air pump ascends, you must hook the chain of the pump brake to a lower part of the pump chain, by which means you can keep pumping until the engine has made its full stroke.

LXX. To make putty for making or repairing the joints. Take whiting, or chalk finely powdered, dry it on an iron plate, or in a ladle, until all the moisture be exhaled; then mix it with raw lintseed oil, and beat or grind it well, adding more oil or whiting, until it be of the consistence of thick paint, and perfectly free from lumps or inequalities.

For some purposes, where the putty is wanted to dry and to be very sticky, use painter's drying oil, which is made by boiling the oil with a small quantity of litharge or red lead.

Where the putty is wanted to continue always soft, mix about two ounces of butter, or common sallet oil with each pound or pint of the lintseed oil: This soft putty is principally useful in the caulked joints of the eduction pipe, above water. N.B. White lead will not answer in place of the whiting.

No wet cloaths should be suffered to be laid on the cylinder, boiler or steam pipes, and every part containing steam should be guarded as much as possible from the influence of cold air or water.

The proper grease for the piston and cylinder stuffing box is melted tallow, and for the chains, gudgeons, &c. common Spanish olive oil (called sallet oil) which for some uses may be thickened by dissolving some tallow or butter in it, by means of heat. – Lintseed oil should never be used as grease, as it dries and creates more friction than would have been without it. – Hogs lard, or train oil, if applied any where about the cylinder, or where it is hot, will thicken like lintseed oil. – When the oil or grease about the great chains, or any of the working parts, grows clotted or very thick, it should be scraped off before new grease is added.

Additional Directions.

The Numbers denote the Paragraph to which they correspond.

VI. As the whole weight of the great beam, and also of the power to be exerted, is supported by the plummer blocks, care must be taken that they stand firmly on the spring beams, and that the latter be well supported from the lever wall. To do which, wherever the building is made of bricks, or of indifferent stone work, form the bottom of the opening, under the beam, of three planks of oak, or of the best deal, six or eight inches thick, and twelve or fourteen inches wide. These planks must reach at least four feet into the walls at each side of the opening, one of them must be laid in the line of the outside of the wall, another in the line of the inside of the wall, and the third, which should be the strongest, in the middle, right under the gudgeon. Upon these planks, at each side of the opening place three others of the same dimensions upright; let their upper ends reach to the upper-side of the spring beams, and let the spring beams be let into the uprights, so that only two inches of their thickness shall project beyond the face of the spring beams, and that the remaining four inches of the thickness of the uprights shall form a shoulder under the spring beams, which will support them firmly under the sides which are next the beam, where it is most necessary; for were the insides of the spring beams or plummer blocks to give way to the pressure, and the outsides to be supported, the gudgeon would rest on its points, and by the leverage it would gain thereby, might be broken. The lower ends of these six uprights may have small tenants to fit mortices in the sills, which will prevent their slipping.

IX. *Page 3.* The holes through the great beam for the screw bolts of the martingale tails should be quite easy for them, otherwise the screws will be broken if the logs of the beam come to slide upon one another. The keys to prevent the logs from sliding upon one another, are best made of pieces of very dry and hard oak, two inches thick, six or seven inches broad at one end, and four or five inches broad at the other end; their length being suited to the thickness of the beam.

XVI. *Page 5.* In large engines, where the condenser pumps are consequently heavy, it is found proper to make the bottom of the condenser cistern of planks five inches thick.

XVIII. *Page 6.* An improvement has lately been made in the covering boiler tops. The setting being built up to nine inches above the flues as usual, a course of horse or cow dung, three inches thick, and well beat, is applied to the boiler top; on the outside of that is

Plates for:
'*Directions for erecting and working the newly-invented Steam Engine*'

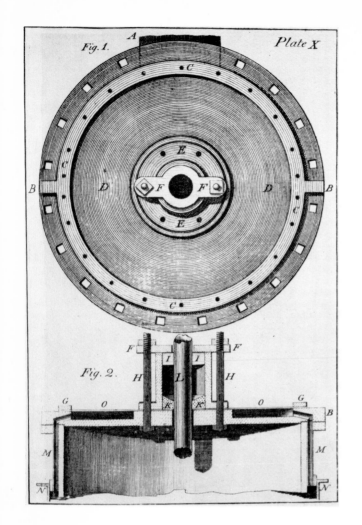

Fig. 1.

Plate X

Fig. 2.

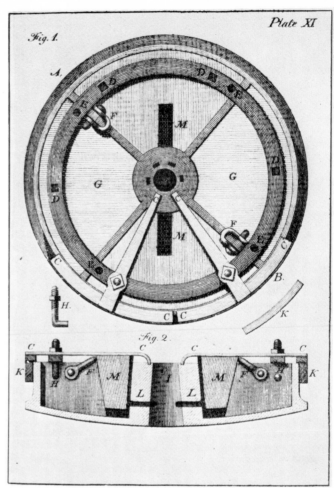

Fig. 1.

Plate XI

Fig. 2.

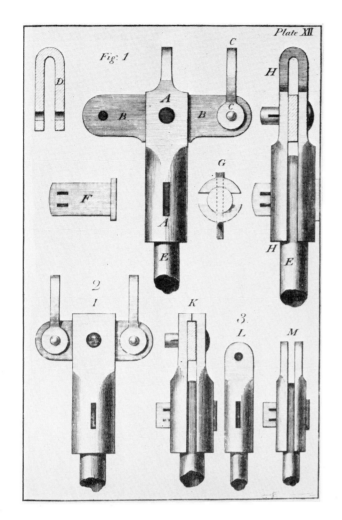

Plate XII

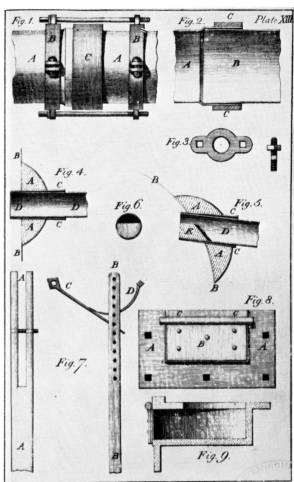

Plate XIII

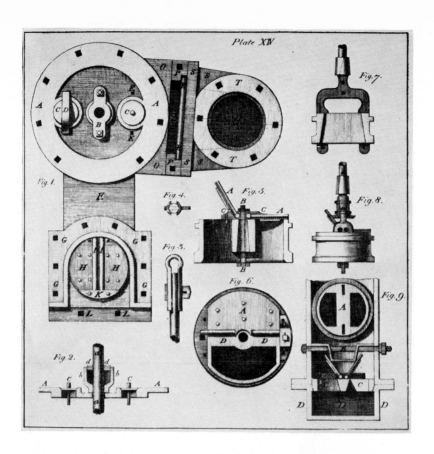

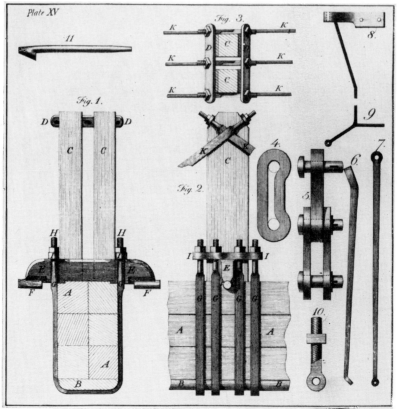

laid some good lime mortar, about an inch in thickness, to which is applied a course of bricks flatwise, with their ends upwards; on the outside of that another course of bricks (also laid in good mortar) in the same position, but so as to break joint with the first course; in which manner the covering is carried on until the whole top is covered, taking care to leave an opening for the man hole; every flanch may be thus covered, and when well done, it effectually makes the top steam tight, and also defends it from cold and rain, so that a boiler house is not necessary. The mortar employed must be such as stands water.

XIX. *Page 6.* The valve put into the boiler feed-pipe, to prevent boiling over, is best fixt in its upper end, so that it may be taken out when any material is wanted to be introduced into the boiler by the steam pipe. The proper valve for this purpose is one of the kind which are used for the injection, and blowing pipe, which must be put into the feeding pipe, in an inverted position.

XXIII. *Page 7.* Instead of using painters drying oil to make the joints with, take good raw or unboiled lintseed oil, put it in an iron pot, place it over a gentle fire, (out of doors, but protected from rain) let it be watched as it heats, as it is very liable to boil over; when it boils make the fire more moderate, but continue to heat the oil, until upon dropping some of it upon a cold stone or piece of iron, you find it is, when cold, of the thickness of thick tar or treacle. The pasteboards for the joints are to be soaked in this oil warm, or painted over with it, and laid in a hot place to suck it up; and it is also to be used to make the putty with.

XXVII. *Page 8.* Instead of putting a gland across the bottom of the piston rod, to prevent it from dropping, it is better to drill two opposite holes through the cone of the piston, and one inch each into the cone of the rod; two iron pins put into these holes will effectually keep the rod in its place. There should be a groove about a quarter of an inch deep, and half an inch wide, cut round the base of the cone on the rod below these pins, which grooves being lapped round with rope yarn and putty, will serve to prevent steam from getting through the piston by the sides of the pins. To make it more easy to get these pins out, they should have flat tails bent upwards, so as to lie close against the outside of the cone of the piston when the pins are in their places; and to secure them there, mortices must be cut in the wood which fills the hollows of the piston, to which must be fitted wooden wedges, made *very* tapering, by driving which down, the tails of the pins will be prest against the cone, and the tapering form of the wedges will make it easy to dislodge them when the pins are wanted to be taken out. It is necessary to observe, that the pins should be fitted tight into the holes in the piston cone, and that the holes into which their points enter in the cone of the rod, should be made easy for them, otherwise they might prevent the one cone from being pulled far enough into the other.

XXIX. *Page 8.* The oakum with which the joints are caulked, should be well smeared with the strong or thick boiled oil, mentioned in these additional directions. If the under side of the pipe of the inner bottom does not fit close to the lower edge of the opening made for it in the outer bottom, that is to say, if the space left there for pasteboard or caulking be wider than one quarter of an inch, a piece of hammered iron an inch and an

L

half broad must be forged of such thickness as to fill up the space so as to make it tight by the help of a thickness of pasteboard above it, and another below it. Lead ought not to be used in these cases, as its expansion and contraction by heat and cold, are too great. Instead of putting a prop from the nozle to the ground, it is found better to put a balance beam off sideways under the floor, with a short upright having a flat end to take a broad bearing under the nozle. The weight of the balance should not support more than two thirds of the weight of the nozle.

XXXI. *Page 9.* The lower ring on the cylinder, to which the steam case is fixed, is sometimes made with a projecting flanch, on which the steam case rests, and the joint is then made tight by caulking between the flanch of the steam case, and that on the ring.

XXXII. To avoid the inconvenience of the perpendicular steam pipes occasionally proving too short, they are now made without any flanch at the lower end, and a socket is cast upon the nozle to fit them, in which they are to be made tight by caulking.

The weight of the upper nozle must be supported by a prop from the cross piece between the cylinder beams. And if the boiler steam pipe be very long, and consequently heavy, part of its weight should be supported by a balance beam near the wall of the house.

XXXV. *Page 10.* The best way of making the standing joints of the condenser, is by means of rings of lead a quarter of an inch thick, as broad as the flanches, and pierced for all the screws. They may either have putty made with the thick oil put on each side of them, or, for greater security, they may be covered with Russia duck and putty. In other respects proceed as directed in xxxv. The soft rope does not answer well.

XXXVI. *Page 10.* In addition to this article, see the explanation of plate xiv.

XXXVIII. *Page 11.* Where the joints of the eduction pipe are made with flanches, they are to be fixed together by means of strong flat rings of iron, put on each side of them, as has been directed for that at the nozle and the joints are to be made tight by pasteboard and putty; for, on account of its expansion, lead will not answer where it is subjected to be alternately hot and cold.

XXXIX. *Page 12.* The hot water pump must be fixed down by means of two long bars of iron with screwed ends, which go through the bottom of the cistern, and extend upwards through two of the holes of the lower flanch of the hot water pump. One of these bars is shown in the drawing of the front of the engine house.

XLII. *Page 13.* The guide posts may be fixed upon a sill passing from one to the other; and the best way of fixing the weight of the exhaustion regulator is to make it in the form of a saddle, moveable at discretion, upon a beam centred at the further guide post, so that the beam may fall flat upon the sill when at lowest, and the saddle will produce the effect of a greater or lesser weight, according as you place it farther from the centre or nearer to it.

The door of the condenser may be converted into a window, and a seat for the engine man, as soon as the condenser and eduction pipe are fixed.

XLV. Some people use a plaited rope to make the joint of the cylinder lid, which is a bad practice; for though a plaited rope may make a joint apparently steam tight, yet it has been found by experience, that such joints are not air tight; but when, by the working of the top regulator, a partial vacuum is produced in the upper part of the cylinder; they permit some air to enter imperceptibly, and without noise, which in course passes to the condenser; and by persons that are not aware of this circumstance, may be thought to enter at some air leak in another place. *We therefore recommend that this joint be always made with pasteboard and putty; and that a strict attention be paid to the tightness of the stuffing box, wherever the top regulator is used.*

XLVI. *Page 23.* The proper quantity of tallow to grease the piston, is two pounds every week for every foot the cylinder is in diameter. But where opportunity can be obtained of adding it more frequently, the whole quantity ought not to be added at once, but divided according to you opportunities. When the top regulator is used, if the tallow is put into a flat funnel which ought to be made to surround the piston rod above the cylinder stuffing box, it will be gradually sucked in without the trouble of taking off the lid.

DIRECTIONS FOR WORKING ROTATIVE ENGINES c. 1784

Directions Relating to the Engine

I. Everything to be kept as clean as possible.

II. When the parts are oiled or greased, the old scurf must be taken off as much as possible before the fresh grease or oil is applied.

III. The working gear ⎫
IV. The parallel motion ⎬ Should be oiled every twelve hours of working.
V. The rotative motion ⎭

VI. The regulator valves should be examined and cleaned once a week, if the Engine goes night and day; or once a fortnight if it goes in the day time only.

VII. Care must be taken to put the valves in with the same side towards you, as they were taken out with.

VIII. The spindles of the valves should be examined when the valves are, and the lapping repaired if necessary.

IX. The piston should be packt once a week, and *seventy*[1] ounces of tallow melted each time, and poured upon the packing before the upper gasket is put in; but if the packing be wholly new, as much tallow should be put in as will soak up; and so much packing should be put in as will keep the under side of the ring from touching the upper side of the piston, when the ring is screwed down.

X. When the piston is packt, the sides of the top part of the cylinder should be scraped to take off the old grease, putty, dirt, &c. that is generally retained there; and the greatest care should be taken that there remains no dirt of any sort, either on the piston, or in the passage to the upper valve.

XI. In packing the stuffing box of the cylinder, so much packing must be put in as to prevent the gland from touching the top of the stuffing box; otherwise there will be no

---

[1] These words are written by hand to suit the particular size of engine involved.

certainty that the box is tight; in which case, either air will pass through it, and injure the vacuum, or much tallow will be wasted, and will not cure the evil.

XII. About 2 *ounces* of tallow may be put in the cup of the stuffing box once in six hours, which will be enough, when the box is tight, to keep the packing elastic.

XIII. The air pump bucket should be examined every month, and new packt if it require it.

XIV. The brasses of the parallel motion should be frequently examined, and the cutters driven up *with a small hammer*, two or three times a day; as any shake there causes noise and bad work, and *vibrations* in the millwork; and the same attention should be paid to the brasses of the connecting link, for the same reason.

XV. The brasses in the connecting link must be packt up with thin bits of iron or copper, from time to time as they wear, so as to prevent the teeth from bottoming, or the edges of the rims upon the sides of the wheels from touching in any part of their revolution.

XVI. The brasses at the top of the connecting rod should be examined, and any shake that may arise prevented by cuttering them up.

XVII. The sliders that shut the hands of the working gear, should be so adjusted as just to let the catch slip into its place at each end of the stroke; and the catch should be pressed by a *weak* spring to make it act quickly.

XVIII. The catch should be so trimmed as to disengage as finely as possible, that is as near the end of the stroke as can be: Much of the smoothness of going depends upon that circumstance.

XIX. The boiler should be cleaned at least once a month; but if the water be muddy or scurfy, more frequently; as it will otherwise not only be liable to destruction by burning, but will likewise require more coals: *Two evils to be carefully guarded against.*

XX. The *water* in the *boiler* should be kept as *nearly* of the *same height* as possible; as carelessness in this point may cause the most sudden destruction of the boiler, and the consequent stoppage of the works (where there is not a spare one ready to supply its place).

XXI. The flues round the boiler should be cleaned once in six weeks.

XXII. The coals when of that sort which does not cake together, should not be of the small kind, but of the lumpy, and the quantity thrown in at a time should be rather small than much, and more frequent the times.

XXIII. The coals, if they do not cake, should not be heaped upon the grate, but should be of an uniform thickness, 6 or 8 inches thick.

XXIV. The weight upon the safety valve should never be so much as to raise the index in the steam gage, in any case, more than three proper inches, or 6 upon the gage, as too great a strain will be brought upon the boiler unless this circumstance be attended to.

154

33. 'List of improvements chiefly mechanical not secured by patent.'
MS. 24 September 1796[?], Doldowlod.

Watt introduced many technical improvements in the construction of his engines which he did not patent, mainly because they were of relatively minor importance, compared with his main inventions, or because they had previously been used in a similar form in other types of machinery (as in the case of the governor, based on windmill experience).

Suspension of the Piston on a conical part of the Rod fitting into a hole of the same shape in the Piston whereby it was held firmer than by the former more imperfect methods
   Forming the piston and serving the packing in a better manner
   Making all the working parts of the Engine with greater Accuracy
   Application of improved valves in place of the old sliding regulators with better means of opening and shutting these Valves
   Various improvements in Working Gear
   Suspending the Beam so that the center of motion was below the center of gravity instead of above as in the Old Engines
   Governor for regulating the speed of the Engine
   Improvement in Boilers and Grates particularly the feeding Apparatus for keeping the boiler regularly supplied with water –

34. James Watt junior to F. Arago,
13 October 1834, Copy, Doldowlod.

It was inevitable that an invention of such major importance as Watt's steam engine should soon be 'pirated' by other engineers, particularly by those who had acquired technical knowledge and experience in Boulton and Watt's own employment. We have already noticed the early unpleasantness with Hateley, then the piracy of the crank by Wasbrough and Pickard, and the beginnings of more extensive piracies by such men as Hornblower and Bull. For a long time, however, Boulton and Watt refrained from legal action, partly through fear of having their Parliamentary monopoly challenged in the law courts, partly through confidence in the technical and economic superiority of their own engines. But eventually in the early 1790s, when these piracies became more numerous, blatant and threatening, they were driven into a series of legal actions, which went on continuously and confusingly until the expiry of their monopoly in 1800. The course of these complicated legal struggles is briefly and conveniently summarized in this letter many years later.

'My dear Sir

'Resuming the subject with which I concluded my last letter, I proceed to state that my father's Improvements on the Steam Engine were introduced into use about the year 1776, or 1777, and their advantages were no sooner proved and acknowledged, than attempts were made to evade the patent by variations in the construction, retaining the essential principles. The most important of these were made by a person of the name of Jonathan Hornblower, who had long been established as an Engine maker in Cornwall and was employed by Boulton and Watt to erect some of theirs. In July 1781 he took out a Patent for "a Machine, or Engine for raising Water by means of Fire and Steam", the Specification of which is dated 5 November 1781. The leading deviation from my father's, consisted in employing a second and larger Cylinder for the Steam to expand in and operate a second time after it had done its office in the first Cylinder, and having a division plate or diaphragm in the large Cylinder near the bottom, under which the Steam was condensed, or sometimes blown out by bellows into another Vessel. This formed a complex and expensive and not so efficacious a Machine as the common construction of Mr. Watt, who expanded the Steam in a single Cylinder, which construction he constantly practised. Few of them were therefore made, but on Hornblower endeavouring to obtain an Act of Parliament in 1792, he was opposed by Boulton and Watt, and his Bill thrown out upon the second reading, on the evidence of Mr. Rennie, proving the infringement and the inferiority of the Machine, by a Majority of 63 to 20; nearly all the Cornish Members present voting for Hornblower.

'It was intended to have proceeded at Law against the Engines he had erected; but another person, of the name of Edward Bull, who had also been employed by Boulton and Watt in the erection of their Engines, having in the meantime made other Steam Engines with the Cylinder inverted (which Mr. Watt had done before) and using the other Inventions of Mr. Watt, it was judged more expedient to proceed in the first instance against him, as his formed a cheaper and better Engine than Hornblowers and was spreading fast.

'An action was accordingly brought against Bull, and the Trial took place in the Common Pleas, in London, in June 1793, before the Chief Justice Eyre and a Special Jury. Mr. Serjeant Adair was leading Counsel for Boulton and Watt. The Witnesses examined on their part, were, Mr. J. A. de Luc, Dr. Herschel, Dr. Lind, Mr. Robert Mylne, Mr. Samuel More the Secretary of the Society of Arts, Mr. Ramsden the Mathematical Instrument Maker, Mr. Alexander Cumming a celebrated Mechanist, Mr. John Southern an Engineer, to prove the originality and advantages of Mr. Watt's Invention and Mr. William Murdock (who afterwards discovered the economical application of Gas lights) Mr. Richard Mitchell and Mr. Peter Godfrey to prove the infringement by Bull.

'The defendants case was conducted by Mr. Serjeant Le Blanc, who called for Witnesses, Mr. Sansom a clockmaker, and Mr. Bramah, Mr. Jabez Hornblower (brother of Jonathan) Mr. John and Mr. William Braithwaite, Mr. Thomas Rowntree and Mr. Richard Trevithick, who all called themselves Engine makers. After their evidence had been given,

Serjeant Adair rose to reply, when the Foreman of the Jury stopped him, saying "they were perfectly satisfied".

'Mr. Justice Eyre then said, that although this was proved to be a noble Improvement and every part of it new, and the Piracy manifest, and the Jury quite warranted to find for the Plaintiffs; there was still a question behind, viz. whether the Specification was sufficient in point of law, whether it was such a description of an organized machine as the Law required, which Question he should reserve for a Special Case.

'Such Special Case was subsequently agreed upon, and will be found in the Pamphlet of the Judges Arguments page 16 to 19. It afforded ample scope for the Ingenuity of the Lawyers, and after two hearings before the four judges of the Common Pleas, came on for Judgement on the 16th May 1795: when two of the Judges, Mr. Justice Rooke and the Chief Justice Eyre who had raised the Question, gave judgement in favour of the Verdict; and the other two, Mr. Justice Heath and Mr. Justice Buller, against it. In consequence of which, the Case was suspended.

'During the interval between the Trial and the delivery of the Opinions of the Judges of the Common Pleas, the Court of Chancery had granted Injunctions founded on the Verdict of the Jury, restraining farther piracies of Mr. Watt's patent until the decision of the Common Pleas took place. And when the Case became suspended they still continued to do so, but with an intimation that Boulton and Watt should proceed to another Trial in a Court of Law.

'An occasion for this was soon presented by some Engines made by Hornblower (Jabez) and Maberley a wealthy Currier in the City, who had purchased a patent of a person of the name of Mainwaring for a Steam Engine on a very bad construction, which he employed Jabez Hornblower to make work, and the latter succeeded by using Mr. Watt's inventions. Injunctions were obtained against them, and an Action brought in the Common Pleas, where it was tried in December 1796, before the same Chief Justice Eyre and a Special Jury.

'Serjeant Adair again conducted Boulton and Watt's case, assisted by other eminent Counsel. The Witnesses for it, were Mr. J. A. de Luc, Dr. Herschell, Mr. Ramsden, Dr. Robison, Mr. Alexander Cumming, Mr. John Southern, Mr. Robert Mylne, who all proved the originality and utility of Mr. Watt's Invention and the sufficiency of the Specification, and Mr. Barnes and Mr. Smalley practical Engineers, to prove the same and the Infringement. Mr. Serjeant Le Blanc again conducted the defence and the Witnesses were, Mr. Thomas Hateley an Engine maker, Mr. Thomas Simpson Engineer to the Chelsea Water Works, Mr. Bramah Engine maker, Mr. Davies Giddy (now Sir Davies Gilbert) and Mr. William Nicholson author of the Chemical Dictionary, their whole evidence going merely to the insufficiency of the Specification. Mr. Serjeant Adair made an able reply, in which he distinguished between the evidence of Messrs. Giddy and Nicholson, and of the Engine makers, who he observed came there "to prostitute their own ignorance". The Chief Justice having summed up, the Foreman of the Jury immediately said, "We find for the Plaintiffs". This Verdict was confirmed by the Court, it being understood that the defendants would carry it by writ of Error to the Kings Bench.

'They did so; and after two long and most elaborate Arguments, in which Mr. Rous, the Counsel for Boulton and Watt, particularly exerted himself, the cause was finally decided in January 1799 by the unanimous Judgement of the four Judges, Lord Kenyon Chief Justice, Sir William Ashurst, Sir Nash Grose, and Sir Saulden Lawrence, affirming the Judgement of the Court of Common Pleas.

'Thus ended this long, harassing and expensive contest; and Boulton and Watt immediately proceeded to recover the arrears of Savings due to them, both from those who had infringed their Patent, and from many whose Engines they had themselves made, but who had suspended their payments until the decisions of the Courts of Law were established. To those who had continued to pay, they made large and liberal allowances, and they exacted far short of their dues from their opponents. The Act of Parliament expired in 1800 and Mr. Watt then retired from business.

'If it should upon reconsideration appear to me that I have omitted any thing material, I shall endeavour to supply it in another letter, although I presume you have by this time had enough of Law.

> I am
> my dear Sir
> Yours truly
> JAMES WATT'

35. 'Mr. Hornblower's Case Relative to a Petition to Parliament for the Extention of the Term of his Patent.' *Printed. 1792. Doldowlod.*

Boulton and Watt had permitted Jonathan Hornblower to patent a double-cylindered, high-pressure engine in 1781, though it certainly pirated most of the principles of Watt's engine. Hornblower did not, in fact, present a very serious challenge for some years, but when, in the early 1790s, he began to achieve greater technical success, posing a threat to their Cornish trade, and tried to secure a parliamentary extension of his patent, Boulton and Watt at last decided to take action.

At a very early Period of Mr. Hornblower's Life, he conceived an Idea of improving the STEAM ENGINE; and in 1776 (previous to his Knowledge of any other Kind than those invented by Mr. Newcomen), made a small working Model; the Effect of which he exhibited to a confidential Friend.

The Principle he sat out upon, was perfectly new, and consisted in a secondary

Application of the Steam, to produce a new Action by the Intervention of a second Steam Vessel.

In all other Machines of the Kind, it was, and is usual, after one Operation of the Steam on the Piston, to condense or destroy it.

The general Principles and Advantages of Mr. Hornblower's Engine must therefore appear obvious to every Person who considers, that whatever Effect can be produced by a new Action of the Steam on a second Piston, must have been lost in all other Steam Engines. After Mr. Hornblower found that the Principle of his Invention could be applied to Practice, he went on to mature it, by several other Models of larger Dimensions; and whilst pursuing his Purposes Mr. James Watt came into Cornwall, by whose Improvement on Steam Engines the very great Advantage of Mr. Hornblower's, over the old Sort, was considerably lessened. He however perceived that his Invention was much superior to Mr. Watt's, and the Nature of the two Improvements totally distinct from each other: His Opinion he also further confirmed by a Comparison between Models of each Engine, on the same Scale.

On Account of some Family Occurrences Mr. Hornblower did not apply for a Patent till 1781; after which he exhibited his Scheme the first Time for Practical Use to the Public, on a Coal Mine in Somersetshire.

This Engine did not succeed at once; as might naturally be expected, though, after some Alterations in the Machinery, it produced an extraordinary Effect. But the popular Opinion of his Machine was fixed and determined, not from the ensuing Success, but from the first Failure in attempting to render it effective; and thus the Principle of the Engine was condemned for a Defect, which existed only in the Mode of applying it. This Circumstance concurring with the Fears of the Public from a Pretended Claim of Messrs. Boulton and Watt, rendered abortive all future Attempts to establish it, till the Year 1790; when an Engine was erected on Tin-Croft Mine, in Cornwall, which has been attended with the greatest Success, and has been admired by all those who have a Knowledge of its Principles, and have attended properly and impartially to the Effect of its Produces [*sic*]. During the above Lapse of Time, Messrs. Boulton and Watt held out threats of a Prosecution, and tired the Patience of the Public, with waiting to see them put into Execution. This Procedure had the Effect of wasting away the Time of Mr. Hornblower's Patent, when a Decision would have enabled him to reap the Fruits of his Labour, and reimburse the heavy Expence attendant on such important and complicated Business.

Mr. Hornblower does not pray that he may have the Benefit of his Patent for more than the usual Term, though Mr. Watt's Grant for a Term of Thirty Years is an Example for him; but he only petitions that such Time as has been lost in great Expence and Disappointment may be added to his original Term, so that he may be enabled to secure such Profits as were intended when he obtained the Grant of his Patent.

With Respect to the Claim of Messrs. Boulton and Watt, Mr. Hornblower cannot conceive that a mere Extension of his Term can in the least impower him to withhold from them what they consider as their exclusive Right, and therefore it will be a peculiar Hardship on Mr. Hornblower to be crushed by the mere Power and Opulence of his

159

Opponents. Mr. Hornblower not having yet made very large Engines, has it not in his Power to compare, on fair Grounds, his Machine with some very powerful ones of Messrs. Boulton and Watt; therefore the latter may seem to lessen the Merit of Mr. Hornblower's Invention; but the only possible Way that a just Comparison can be made, must be between Engines of the same Magnitude; because the Advantages in Favour of large Engines are on various Accounts, very great.

Mr. Hornblower has only further to remark, that his Confidence of Success originates only from the Justness of his Cause, and the Reasonableness of his Petition, as not calculated to abridge the Rights of any Person; and that he may be rewarded, for the best Part of his Life having been spent in the Pursuit of Mechanical Improvement: that his Expectations of Pecuniary Assistance from his Patent, has induced him to lay himself out in such a Manner for that Purpose, as will be the Cause of his sustaining very considerable Injury with Respect to his Family's Support; and for ever damp his Ardour for further Improvement, if his Petition be not granted.

36. 'Short Statement, on the Part of Messrs. Boulton and Watt, in Opposition to Mr. Jonathan Hornblower's Application to Parliament for an Act to prolong the Term of his Patent.' *Printed. 1792. Doldowlod.*

The Steam Engine, invented by the Marquis of Worcester, was brought into Use, in an imperfect State, by Captain Savery. Mr. Newcomen added to it a Piston, moving in a Cylinder, and working a detached Pump; and thereby greatly changed and essentially improved, it's Mode of Action, as well as reduced it's Consumption of Fuel. Yet it still required so great a Supply of Fire, that many principal Mines in Cornwall, where Coals are dear, were upon the Point of being abandoned, and several had been so, when Mr. Watt opportunely invented his Improvements in the Construction of the Engine. It is well known that those Improvements, by their great Saving of Fuel, have re-instated the Cornish Mines, and, through the Exertions of Mr. Watt and his Partner Mr. Boulton, have been brought into very extensive Use throughout the Kingdom, and successfully applied to a great Variety of Purposes highly important to the National Industry and Prosperity.

Mr. Watt, in 1769, obtained a Patent for his Invention; and in 1775, an Act of Parliament was passed, whereby his Interest in the Invention was secured to him for Twenty-five Years from that Period.

[Here follows an account of Watt's engine contrasting it with Newcomen's.]

In 1781 Mr. Jonathan Hornblower obtained a Patent for a Fire Engine, alledged to be of his Invention.

If there had been Merit and Novelty in that Engine, Messrs. Boulton and Watt would have readily acknowledged the Facts. – They were willing that Mr. Hornblower should bring his Engine to a fair Trial, and therefore have not hitherto interrupted him: But now, after *Eleven Years Trial*, Mr. Hornblower having erected only Two Engines, and those *upon the same Principles* as Mr. Watt's, in all essential Points; and attempting, as he is now doing, to procure the Sanction of the Legislature to his Proceedings, it becomes impossible for Messrs. Boulton and Watt longer to continue silent Spectators, or to permit an Imposition to be practised so injurious to their just Rights. They, therefore, undertake to prove, by competent Evidence, *that Mr. Hornblower's Engine is a direct and palpable* PLAGIARISM *of Mr. Watt's Invention.*

With this View Boulton and Watt propose to prove,

1. That Mr. Hornblower makes use of the expansive Force of Steam to press down the Pistons of his Cylinders, as Mr. Watt does, although he (Mr. Hornblower) has omitted to state in his Specification what acting Power he proposed to use for that Purpose.

2. That in Regard to the 'Condensation of the Steam, by causing it to pass in Contact with Metalline Surfaces, while Water is applied to the opposite side', as expressed by Mr. Hornblower in his *Specification*; it is only one of the many possible Ways of using Mr. Watt's Method, set forth in *his* Specification, of condensing by 'Application of Water, or other Cold Bodies, to the Condenser', and was publickly used by Mr. Watt, prior to the Year, 1775.

3. That, *in Practice*, Mr. Hornblower does not condense or destroy the Steam in the cylinder itself, *but in a Place distinct from it.*

4. That Mr. Hornblower does not use the Means expressed in his Specification, to discharge the Engine of the Water, Air, and condensed Vapour; but, in Fact, employs the very Means pointed out and employed by Mr. Watt, namely, a Pump wrought by the Engine.

5. That Mr. Hornblower makes use of Oil, Wax, or other similar Substances, to make his Pistons Air-tight, in the same Manner as expressed in Mr. Watt's Specification, and as used by Boulton and Watt.

6. That the Two Engines constructed by Mr. Hornblower, are (notwithstanding his having made use of Mr. Watt's Inventions) not superior in any Respect to those erected by Boulton and Watt, but are really inferior; – *contrary to the Allegations contained in his Petition and Bill*!

When these Points are proved, Messrs. Boulton and Watt trust, that neither the Legislature, nor any Individual, will think Mr. Hornblower's Pretensions intitled to any Countenance or Favour.

37. James Watt to Thomas Wilson,
18 July 1792, Truro.

In addition to baulking Hornblower in this way, Boulton and Watt also tried to overcome such rivalry by demonstrating to the Cornish miners the superior technical efficiency and economy of their own engines.

'In regard to the publication, I think it will be right to publish the savings which have been made by our Engines over the common ones at the Mines you mention and from each of these drawing the inference of what coals would have been consumed if the Common Engine had been of the same power as to going deeper that ours were, after which you may slightly touch upon the total amount that would have been saved or was saved on *all* the Mines you mention, but I think the comparison should be fair and depths admitted as well as gallons. You may then shew the comparative effects of our Engines and Hornblowers from which you may draw the inference that $\frac{1}{3}$ or more of the fuel is saved by our Engines when compared with Tin Croft in similar circumstances, but I would offer no undertakings. The facts are sufficient to any one who reasons and the offer to undertake would imply some right in our opponents, which we ought by no means to allow to be implied, and in the whole of the paper you must confine yourself to mere facts and the natural inferences from them, without saying any thing either personal of the Horners or of their rights which you profess at present not to discuss only to disabuse the county as to matters of fact.

'In respect to Hallamin if they will not execute an agreement the same as others, we cannot come into their terms, and may as well try the matter now as after establishing such a precedent for if that Engine is to work both Hallamin and Rettallack it cannot work them long and then we shall be at 6s and 7s again; however our former proposal may be repeated to them that what further power they may want to work these two mines we shall charge no premium upon – If you can keep them off and on for a few days it may be well as I am going to London to take some advice and we should act accordingly and the same doctrine may be applied to Poldice, a positive answer to which should be declined, at any rate until you have some better authority for the offer than Mr. Vivian's talk. If you do not get proper information from Tincroft as to effects and consumption you may state it so in your paper, but mention no names rather excusing the Captains as acting under the direction of the Adventurers and consequently in the way of their duty – You say you wish we were as determined as you are, we are determined to do whatever seems most for our advantage once we can see clearly what that is; but it is truly said that law is a bottomless pit and we would fain see some good firm ladder to get out before we venture to go into it. The Justice of our cause we cannot entertain a doubt of nor does any body, but the labour, the uncertainty, and the expence of a suit are not light things. Yet those we shall encounter rather than suffer the very gross injustice with which we are treated and we request you to furnish us with such materials and information as you can respecting Poldice and Hallamin.'

38. Thomas Wilson, 'An Address to the Mining Interest of Cornwall, on the subject of Messrs. Boulton & Watt's and Mr. Hornblower's Engines.'
Printed. 1793. Doldowlod.

To
The Lords of,
and
Adventurers
in

THE MINES OF CORNWALL.

GENTLEMEN,

So much has been already said respecting the Engine erected by Mr. HORNBLOWER, at Tin Croft Mine, and its effects when compared with the Engines erected by Messrs. BOULTON and WATT, that I would not enter into any further discussion on that head, except to clear up some points which seem not to be well understood, and which have industriously been mistated, to persuade the public into an opinion of the merits of that Engine, which it by no means has any title to.

These points are, Mr. Hornblower's assertion in a paper delivered to the Members of the House of Commons, 'That his machine, on a just comparison with Mr. W.'s, is found to be as 16 to 10 superior in its effects.'

Messrs. Hornblower and Winwood's Advertisement of the 30th July, 1791, in which they say, 'they have by their Engine at Tin Croft, exhibited a Machine which evidently surpasses every other one of the kind *in a double proportion*.'

A friend or advocate of Mr. Hornblower, under the signature of a Disinterested Cornish Miner, has published in the Sherborne Mercury, dated November 8, 1792, a long paper, replete with false assertions and erroneous reasoning, in which he says, 'that 'Mr. Hornblower has been able by an ingenious contrivance, to apply the Steam a second 'time, so as to gain an additional or new power equal to 6 *lbs*. upon every square inch 'of the Cylinder, which being added to the power of its primary action, produces an 'effect of more than 16 *lbs*. to the inch,' and adds, 'that the primary action of the steam 'in Hornblower's Machine, is equal to the power of the same steam in Boulton's, and its 'secondary application gains a new power of about 6 *lbs*. to the inch.' He says further, that Messrs. Hornblower and Winwood will undertake 'to produce the same effects 'with three bushels of coals, as is done with five bushels in Mr. Boulton's Engines; and

163

'that when Mr. Boulton's Engines are got to the extent of their power, Mr. Hornblower's 'Engine is loaded only at two thirds of its burthen, the cheapest period of its working, 'and in consequence the Mine can be pushed one third deeper, without the amazing 'expense of a new Engine.' Other assertions of the like nature are added, which I shall not at present bestow any attention upon, because they depend upon what has been already quoted.

One principal point to be attended to, is the very extraordinary manner in which Mr. Hornblower chuses to state the power of his Engine, and which, If I am to follow the dictates of that portion of understanding which Providence has bestowed upon me, I must maintain to be to the highest degree erroneous, and calculated to mislead those who do not take the trouble of thinking upon the subject.

I shall therefore now state what *I believe* to be the size, the power, and the effects of the Engine at Tin Croft, about which so much has been said more than the subject merited. I have seen the Engine repeatedly, and have been informed of particulars which did not fall under my own observation by the Captains and some of the Adventurers. If I have fallen into any error (as to the facts) I am desirous of being corrected, but I am persuaded there is none of any consequence.

The Engine at Tin Croft has two cylinders, whose pistons act upon the same end of the working beam, the small cylinder is 21 inches diameter, with a stroke of 6 feet long, and the larger one is 27 inches diameter, with a stroke of 8 feet long, and their conjunct powers are applied to work a 6 feet stroke in the pumps; the weight of the column of water I cannot pretend to speak positively to, as I have never descended into that Mine, or measured the length and diameters of the pumps myself, but believe the work done has been represented more than it ought, as I have reason to doubt whether in Summer particularly, the tye lift, or upper tier of pumps was fully supplied with water, if not, the load will be materially lessened; but as Mr. Hornblower and his friends have uniformly declared that his Engine will work under the load of 16 *lbs.* per square inch, calculating upon the primary or small cylinder only, I will admit it, for the sake of some data to ground my calculations upon, having no doubt but by a little juggling it will work under a much greater load, for in proportion to the *imperfections* of the piston in the small cylinder, the power will be the greater in the large one; insomuch, that if the packing of that piston is taken out, or a hole made through the piston, or it be totally taken away (leaving the rod working to preserve appearances) the Engine will then become similar to B. and W.'s single Engines, and its power increased so as to amount to 23 *lbs.* for every square inch of the area of the small cylinder, although, in fact, there will be no power at all produced in the latter, under those circumstances.

But to proceed to my calculations, I say the area of the 27 inch cylinder, is $572\frac{1}{2}$ square inches; but as its stroke is 8 feet long, and the stroke in the pumps only 6 feet, I must therefore suppose another cylinder to be substituted for it, which would have the same power with a 6 feet stroke as it has with an 8 feet stroke, and the area of that cylinder would be 763,4 square inches, the area of the 21 inch cylinder is 346,36 square inches, which added to the other, makes in all 1109,76 square inches acting with a 6 feet stroke;

this being the effective number of square inches contained in the areas of the two cylinders at Tin Croft. If I follow the usual mode of calculation to find the load per square inch, I must divide the weight of the column of water in the pumps, or the load 5541 *lbs.* by the square inches in the areas of the cylinders, and the quotient will shew 5 *lbs.* nearly to be the load on each square inch of the areas of the cylinders. I leave Mr. Hornblower to reconcile this fact with his pretended load of 16 *lbs.* to the inch. I know he will answer, that he calculated only upon the area of the smaller cylinder; but I must ask what right he has to do so; as at the commencement of the stroke, the smaller cylinder at Tin Croft is full of steam of equal density both above and below the piston, it can have *no power* at the beginning of its motion, that effect being caused solely by the action of the steam upon the piston of the large cylinder, but as the pistons descend, that of the smaller cylinder is acted upon, and its power increases gradually till it is at the *end* of the stroke, when according to the theory assumed by Mr. Hornblower (viz. that the power is inversely as the spaces which the steam is expanded into) the power upon that piston is equal to 5,4 *lbs.* upon each square inch (supposing the full power equal to 10 *lbs.*) and the average power will be found by calculation equal to 3,4 *lbs.* upon the inch through the whole length of the stroke, which is only equivalent to 1,5 *lbs.* upon the inch of the large cylinder. In the large cylinder, as there is supposed to be a vacuum under the piston, the steam will act upon it at the beginning with its full power, equal to 10 *lbs.* per square inch; but as its piston descends, the force upon it will gradually diminish, because the steam from the first cylinder continues to expand into increasing space (for it must not be supposed that any steam can pass by the first piston) till at the bottom of the stroke it is expanded into more than double the space, and according to the theory, must press with less than half the power, viz. 4,5 *lbs.* on the inch, and the average power will be found by calculation to be 6,5 *lbs.* on the inch, adding the average powers of each cylinder together, that is 6,5 and 1,5 will produce a power equal to 8 *lbs.* on the square inch of the large cylinder through the whole stroke, which will be only 5,5 *lbs.* on the square inch, reckoning both cylinders; and this is the full power of the Engine by the most favourable theory.

[N.B. In all the above calculations, when speaking of the great cylinder, its advantage of leverage of 8 to 6, has been considered and included.]

It appears from what I have said, that so far from the greater part of the power, or 10 out of 16, being exerted in the smaller cylinder, its effects according to Mr. Hornblower's assumed theory, can only be equal to 3,4 *lbs.* on the square inch upon the average throughout the whole stroke, equal only to 1,5 *lbs.* on the inch of the larger cylinder, while the power exerted in the large cylinder alone, would be 6,5 *lbs.* on the square inch average throughout the whole length of its stroke, or more than four times the power exerted in the smaller cylinder.

Herein is a specimen, among many others, of the ignorance or artful intention of Mr. Hornblower to deceive, by his stating the small cylinder to be the principal one, and calculating the power of the Engine, as if it was wholly or very principally produced in it, thereby pretending to exhibit the *amazing powers* of his Engine, while, in fact, not one fifth part of the whole power is exerted in the small cylinder; of the two ways to misrepresent the power of Mr. Hornblower's Engine, it would have been more modest to have calculated upon the large cylinder than upon the small one, in which, as I have before said, not one fifth part of the whole power is produced, but the fair and true calculation ought to be made upon the *solid contents* of both his cylinders, and then it will be allowed that one cylinder of Boulton and Watt's Engines, of the same capacity as his two, would act under a load of 20 *lbs.* on the square inch, and more, as experience has proved, whilst his will act with only 5,5 *lbs.* on the square inch, and even this number is *greater* than experience warrants to be practicable, it being the result of calculation, upon a theory known to be false, and which by this time he may have discovered to be so, without knowing why it is so.

But all this reasoning about power per square inch, tends only to perplex and mislead, and I should not have spent any time upon it, except to shew the falsity of their statements.

The essential merits of an Engine are, the being less expensive in its first cost, and its requiring less fuel to do the same quantity of work. In both these points, I assert that Boulton and Watt's Engines have manifestly the advantage, if I may be allowed to understand a simple question in common arithmetic; for though it is asserted that Mr. Hornblower will undertake to produce the same effects with three bushels of coals that is done by five bushels in Boulton and Watt's Engines, I affirm that according to the trials that have been made, the fact is the reverse, viz. that Tin Croft Engine requires more coals to do the same work, than Boulton and Watt's do, in the ratio shewn by my former publications on that subject.

If Tin Croft Engine is capable of having its power to easily increased as the Cornish Miner asserts, I ask why it is now incapable of sinking the Mine any deeper, though, in his language, not loaded two thirds of its power? My answer to that question would be, from what I have said of its present condition, that it is not only at its full power, but overloaded, and as it is now constructed, can do no more.

ONE OF BOULTON AND WATT'S ENGINES, WITH A CYLINDER OF THE SAME SIZE AS THE LARGEST AT TIN CROFT, WOULD WORK THE SAME PUMPS NEARLY THREE TIMES DEEPER, or say under the load of 16623 *lbs.* which is three times the load of Mr. Hornblower's Engine, reckoning 16 *lbs.* to the square inch on his small cylinder.

The truth is notorious, and too well attested, to be disputed by those who have enquired into the matter with candour, though much sophistry hath been employed to explain it away, and to put a false gloss upon it.

That my own belief of what I assert may no longer be called in question, nor myself continue to be branded with the appellation of one, who for his own private interest wishes to mislead the county, though my real intention has been to prevent their being misled by others – I make the following propositions, which, if refused by the other party, will shew their diffidence of what they wish others to believe; and if accepted, will shew which of us is in the right.

As I have before stated the load of Tin Croft Engine at 16 *lbs.* per inch on the small cylinder, is equal to 5541 *lbs.* and it makes about 420, or say 440 strokes of 5 feet 10 inches long in the pumps with each bushel of coal, 5541, x by 5 feet 10 inches equal to 32323, x by 440 strokes per bushel, equal to 14,222,120 *lbs*. one foot high, raised by each bushel of coals.

(‡ 'Though the stroke is before called 6 feet, yet it can make no more than 5 feet 10 inches, which is the length Messrs. H. and W. have used in their advertisement of September 12, 1792, in which they charge me with having, in my former publication, made false assertions and groundless comparisons, without in any one instance shewing wherein. It would be naturally imagined, that persons writing principally to accuse others of mistatements, would be extremely careful to avoid any imputation of the same nature; yet in that same advertisement, in one short account they give of the performance of Tin Croft, they have made the coals consumed to be only 21 instead of 22 bushels (the real average consumption per day in August) and stated the Engine to work eight strokes per minute, when my information on the spot from Mr. Joseph Hornblower, was a *little more* than seven strokes; if Mr. Hornblower will reduce the effects by these data, which are the truth, he will find they will come very near what I have above stated.)

If the Lords and Adventurers shall consent thereto, I offer,

I. To erect one of Boulton and Watt's Engines at my own expence on that Mine, the cylinder of which shall be of the same diameter and length of stroke with only *one* of the two cylinders which compose Tin Croft Engine.

II. I engage that such Engine shall have sufficient power to work under a load equal to *three times* that of Tin Croft Engine, when its smaller cylinder is loaded to 16 *lbs.* upon the square inch: and at its cheapest period of working, it shall raise 50 per cent. more water with each bushel of coals than the aforesaid calculation of 14,222120 *lbs.* weight, one foot high per bushel; and I might safely say much more than that, because I have seen one of Boulton and Watt's Engines of less power raise more than twice as much per bushel of coals, but it would be imprudent to tie myself too closely.

III. I also engage that the said Engine shall have power sufficient to work pumps, the columns of water in which shall weigh three times 5541 *lbs.* with a six feet stroke.

[N.B. Tin Croft smaller cylinder is 21 inches diameter, and its area 346,36 square inches × by 16, is = to 5541 *lbs.*]

IV. The effects of the Engine which I propose to build, shall be judged of by candid and disinterested gentlemen of knowledge and experience in those matters, to be appointed by Mr. Hornblower and myself previously to the erection of the Engine; for I will not run any risk of partial proceedings in a matter of so much consequence.

V. I offer to pay into the hands of a Banker, the sum of £.1500, provided Mr. Hornblower will pay the like sum also into the hands of the same Banker, to be disposed of as follows:

If my Engine performs according to my second and third engagements as aforesaid, the whole sum of £.3000 shall be paid into my hands for the following purposes, viz. 1500 to repay my deposit, £.1200 to repay me for my loss of time and expence in erecting the Engine, and £.300 to be paid by me to the Cornish Infirmary. If, upon the contrary,

M

the Engine shall not perform as I now engage, then £.3000 to become the property of Mr. Hornblower, subject only to his making a gift of £.300 to the Cornish Infirmary or Hospital.

Provided always, that which ever party shall have it in their power to make the gift to the Hospital, it shall be recorded in the books of the institution by whom the gift was made, and upon what account. And provided also, that in any event the materials of the Engine shall remain my property, and at my disposal.

I am well aware that if this offer was to be considered as a wager, it would be said that I was laying the odds, as in case of success, I can only have the second-hand materials of the Engine for my reward; and, on the contrary, if the event shall not prove favourable to me, I should lose above £.2000; but I have such knowledge and confidence in the effects of Messrs. Boulton and Watt's Engines, when properly managed, and so thorough a conviction of the incapacity of Tin Croft Engine to bear a competition with them, when placed in the same circumstances, that I have not the smallest fear of loss in making the offer I now do.

Messrs. Hornblower and Winwood have repeatedly in their publications, accused me of unfair partiality, want of candour, and of attempting to mislead from interested motives; they have even accused me of publishing matters as facts, which were not so, and particularly of depreciating their Engine to their and the county's great loss.

Some of their insults I have at various times repelled, but I think it due to myself to say now, that surely I can with less propriety be accused of self-interest, in defending the cause of others, than Messrs. Hornblower and Winwood can be, in speaking and writing in their own praise, and to the prejudice of others, perhaps not perfectly according to their own convictions, and in doing that no men have shewn less *modesty*. To Messrs. Boulton and Watt I am only an Agent; I have endeavoured to be a faithful one, which cannot be to my discredit; but though I am not like some others, ungrateful for the favours these gentlemen have done me, yet were they capable of urging me to propagate falsehood, or to serve them by dishonourable means, I should certainly resist their requests.

I commenced the defence of Messrs. Boulton and Watt's Engines as a duty, in the full conviction of their merit, and of the injustice done to these gentlemen by Mr. Hornblower's encroachment on their rights, as well as to repel the influence of the very false information given to the principal gentlemen of this county, thereby prejudicing the interests of Messrs. Boulton and Watt, by unjustly depreciating the merits of their Engines; my efforts in the cause of what I think justice, *feeble and inconsistent* as they have been termed by my opponents, have drawn from them so much personality to me, that I have insensibly become a party in the dispute, and as such shall continue to defend the cause I have espoused, so long as I continue to think it a just one.

I am, Gentlemen,
Your very humble Servant,

Thomas Wilson.

Truro, January 5, 1793.

A TABLE

Shewing the Sizes of BOULTON *and* WATT'S *Engines, compared with Engines constructed with two Cylinders, after the plan and proportion of* Tin Croft *Engine, erected by* Mr. HORNBLOWER (*8 feet stroke in Cylinder, 6 feet stroke in pump.*)

			Total load in pounds weight.
No.			
I.	An Engine the size of TIN CROFT Large cylinder, diameter 27 inches Area of 27 inches is Nearly equal to 3 such Engines as Tin Croft	572,555 *x* by 28 =	16031
II.	An Engine the size of WHEAL BUTSON The cylinder is 36 inches diameter Area of ditto 1017,88 *x* by 28 = More power than 5 such Ens. as T. C. & lefs furface		28500
III.	ALE and CAKES Mine New Engine Cylinder is 45 inches diameter Area of ditto 1590,4 *x* by 28 = Equal suface with T. C. and 8 times the power.		44531
IV.	POLDICE New Engine Cylinder is 58 inches diameter Area of a 58 is 2642 *x* by 28 = Equal in power to 13,33 such Engines as T. C.		73976
V.	WH. MAID Engine has a nine Feet Stroke in The cylinder is 63 inches diameter Area of a 63 is 3117,24 *x* by 31½ = Equal in power to 17¾ such Engines as T. C.		98193

An Engine equal in power to No. I. and of the construction of T. C. requires
One cylinder of 36 in. diar. and 1017,88 area *x* 16   =
One ditto   46 ditto   and 1662
‾‾‾‾
     82

} 16286

An Engine equal in power to No. II. and of the construction of T. C. requires
One cylinder of 47 in. diar. and 1734,94 area *x* 16   =
One ditto of   60 ditto   and 2887,4
‾‾‾‾
     107

} 27759

An Engine equal in power to No. III. and of the construction of T. C. requires
One cylinder of 59 in. diar. and 2734   area *x* 16   =
One ditto of   76 ditto   and 4536,4
‾‾‾‾
     135

} 43744

	Total load in pounds weight. Total load
An Engine equal in power to No. IV. and of the construction of T. C. requires One cylinder of 77 in. diar. and 4656 area *x* 16 One ditto of 99 ditto and 7697,5 ——— 176	= 74496
An Engine equal in power to No. V. and of the construction of T. C. requires One cylinder of 88½ in. diar. and 6131,45 area *x* 16 One ditto of 114 ditto and 10207 ——— 202	= 98103
TIN CROFT Engine consists of One cylinder of 21 in. diar. and 346,36 area *x* 16 One do. 8 ft stroke 27 in. diar. and 572,55 area with a six feet stroke in the pump. ——— 48	= 5541

Postscript.

The calculations in the foregoing table, are made upon the supposition, that the length of the strokes in the cylinders of Boulton and Watt's Engines, are eight feet (except Wheal Maid, which is nine feet long) and those in the pumps six feet, which proportions I adopt, because the length of the stroke in the large cylinder of Tin Croft Engine (wherein four fifths of the whole power are produced) is eight feet, and that in the pump six feet.

It will be allowed by most Cornish Miners, because they know it to be fact, that Boulton and Watt's Engines work regularly under the load of 10½ *lbs.* per square inch, upon the descent of the piston, and the same in the ascent, which is equal to 21 *lbs.* per inch, which power being exerted through a stroke of eight feet, will be equal to 28 *lbs.* per square inch, through a stroke in the pump of six feet; and the area of a circle in square inches, whose diameter is equal to that of the cylinder, being multiplied by 28, will give the load that Boulton and Watt's Engine will work under with great regularity.

As Mr. Hornblower asserts, that this Engine will work under a load of 16 *lbs.* on the square inch, calculating upon the area of the small cylinder only, I have therefore calculated upon that load in the foregoing table, and the proportion of both cylinders expressed in the table, are in round numbers, the same as in Tin Croft, or as 21 to 27, and calculated as follows:

To find the size of one of Mr. Hornblower's Engines, that will be equal in power to one of Boulton and Watt's.

I. Divide the load expressed in column belonging to Boulton and Watt's Engine, by 16, and the quotient will be the area of Mr. Hornblower's small cylinder of six feet stroke.

II. They say as the areas of his cylinders at Tin Croft (viz. 346 and 572) are to each other, so will the area of Mr. Hornblower's small cylinder, found as above, be to the area of his large one corresponding to it.

The most expensive parts of Fire Engines are,

THE CYLINDERS.
THE BEAMS.
THE BOILERS.

The *expences* of *cylinders* increase in the proportions of the diameters and lengths, as they grow larger they must also be thicker, and therefore the weight and expence will increase in a greater ratio. The *friction* will also be in proportion to their diameters and lengths of strokes. The *loss of heat* from the outside surface, and the *condensation* of steam within will be also in the like proportion.

View but the table, and compare the magnitudes of the cylinders of the two Engines equal in power, and then figure to yourselves the difference of the *first cost*, of the *friction*, of the *loss of heat and steam*, and of the expence of *oil and tallow*.

Moreover, I believe it *impracticable* to cast and bore cylinders so large as would be necessary to make Engines of equal power to that of Wheal Maid or Poldice, at any expence.

In regard to BEAMS I must observe, that the beams for Boulton and Watt's Engines, need only be of half the strength that is necessary for Engines, whose pistons are acted upon in one direction only, because the load is equally divided, and one half of it is suspended at one end of the beam, and the other half at the other end, and thus the gudgeon only supports a weight equal to the whole load, whereas in Mr. Hornblower's Engines it bears twice the weight of the load, and consequently has twice the friction.

As Boulton and Watt's Engines will raise 50 per cent. more water with each bushel of coal than Tin Croft appears to have done, they certainly require less BOILERS.

To conclude: – I hope I have now proved to the conviction of the unprejudiced,

THAT Mr. Hornblower's method of calculating the power by the area of the *small* cylinder only is erroneous, that only one fifth of the whole power can be produced in that cylinder, and that four fifths are due to the large one.

THAT the 27 inch cylinder *alone*, if employed as one of Boulton and Watt's double Engines, would produce nearly three times the effect that *both* the cylinders at Tin Croft actually do.

THAT by the trials which have been made on several of Boulton and Watt's Engines, the effect produced by the same quantity of fuel is at least one half more than the effect produced in Tin Croft Engine. Every Miner in Cornwall doth know and must allow that Tin Croft Engine is an infringement upon B. and W.'s patent, and is certainly constructed upon their principle, although by the variation it is 50 per cent. worse.

THAT an inspection of the preceding table, and consideration of the machines, will shew that B. and W.'s Engines must, at the same powers, cost very much less money than those of the construction of Tin Croft.

THAT Tin Croft Engine is not only loaded to its full power, but beyond it, is proved by its inability to sink the Mine deeper, or to keep it in fork at the present depth.*

THAT if I did not *most sincerely* believe what I have advanced to be true, I must be a fool who have deceived myself, and not a rogue who wishes to deceive the public; otherwise I should not have risqued so much money on the success of a comparative trial of one of Boulton and Watt's Engines against Tin Croft.

If endeavouring to expose error, and to counteract false and malicious publications *be* a crime, I must plead guilty to the charge; but I hope to be judged with candour, and that the county will in the event see clearly who are their friends, and who have misled them.

39. Extracts from 'Boulton and Watt *versus* Bull. Copy of the Short hand Writer's Notes of the Trial in the Court of Common Pleas. June 1793.'
MS. Doldowlod.

These arguments, however, failed to convince many of the Cornish miners and other manufacturers, who saw the possibility of acquiring cheaper, though perhaps less efficient 'pirated' engines, free from payment of royalties to Boulton and Watt. The latter, therefore, had now finally to resort to the law courts, where Watt's fears of this 'bottomless

* I beg leave to state to the Adventurers of Tin Croft Mine, that had they expended the same money which the Engine now in working cost them, in building one of Boulton and Watt's construction of the size I have proposed, they would at this moment not only have had a machine which would have gone with ease fourteen strokes per minute under the present load, with two thirds of the fuel they now burn, but also capable of sinking three times as deep with the pumps now in use; their Mine by this time would have been much deeper at the same expence, and their produce for several months past much greater; the good effects of this measure would have been felt by them, to an extent not easily calculated, but which may be judged of when compared with their present situation and future prospect, they must now either give up, or work on to great disadvantage and probably loss; till, to use the language of the Cornish Miner, they have been at '*the amazing expence of a new Engine*.'

That I am warranted in stating the above, I appeal to those very respectable persons who attended the late trials at Wheal Butson, whether that Engine did not go thirteen strokes, or even more, per minute, with moderate steam, when loaded to very near 9*lbs.* per square inch; and if this is allowed, surely I may safely assert, that an Engine of the same construction would work even at the rate of 15 under a load of 7½*lbs.* only.

The failure of the late trial at Tin Croft to work with *one* of the two cylinders only, proves the Operator's ignorance of the Steam Engine in general, and that Boulton and Watt's assertion of the construction, in so far as it deviated from their Engines, being a deterioration instead of an improvement, was right.

What is above said to Tin Croft Adventurers, applies equally to all those who have chosen to build or order Messrs. H. and W.'s Engines, and I have not the least doubt but each in their turn will experience the same difficulties and loss as the first users in this county of this pretended improvement.

pit' were to be only too amply justified: indeed, a prolonged struggle now ensued, in which Boulton and Watt had to fight desperately to defend and preserve not merely their patent rights as prolonged by the 1775 Act, but even Watt's claims to inventive originality. This legal battle began with the case brought against Bull.

[Mr. Serjeant Adair, counsel for the plaintiffs, first outlined the principles and the defects of Newcomen's engine and then explained Watt's improvements, stressing particularly the crucial importance of the separate condenser.]

'. . . Then the mode of carrying this into effect when it was discovered was plain and easy as any thing could be imagined – There was nothing to do but to make another Vessel of no matter what shape or Construction whether a Globe or a Pipe or Cylinder or of any other construction was totally immaterial provided it was a vessel that would retain Steam that was all that was necessary. It was therefore only necessary to make another Vessel of any Construction whatever, to let it communicate by a pipe with the vessel in which the rarefied Steam was and to have a valve or regulator or Cock or any other contrivance of a similar kind no matter what, which should open and shut it alternately so as to let the Steam from the Steam vessel into the Condenser when it was intended that it should be condensed and a vacuum produced and to shut it up in its own vessel when it was meant that it should retain its elastic Power. I give Mr. Watt the Inventor no sort of Credit for carrying his Invention into effect because the very instant you hear it was to be condensed in another Vessel all the rest followed to the most ignorant and stupid Person upon Earth every Man the Moment he was told this was to be done was besides the operation of his own common Sense already in possession of all the Means by which it was to be carried into effect . . . the common knowledge of Mechanics and the common sense of Mankind pointed out all that remained to be done after you had fixed the point that it was essential that the Steam should be condensed in a separate Vessel. [Valves, it was pointed out, were already used in Newcomen's engine, and 'there was nothing at all new' in the use of an air pump, though 'the application to the Engine was new'.]

'. . . the particular shape and Mechanical construction of these Vessels is totally immaterial and make no part whatever of the invention (that is perfectly obvious when you see what the invention is) it consists in keeping the Steam Vessel in a Uniform degree of heat – first by protecting it from the external cold and by the performing the condensation in another Vessel – in keeping that Condenser cold – in the occasional communication between these Vessels and the manner of making the Piston Air and Steam tight by the application of other substances instead of Water which did not evaporate with the same Degree of heat.

'The application of the Pump to Pump out the Air and Water accumulating in the Condenser was the necessary consequence of the other and was an application of Mechanicks perfectly known – then you see it did not depend at all upon the Shape of these Vessels the Effect of the Condenser would be exactly the same whatever form it was of – the effect of the Steam Vessel itself would be the same if it was a Square an Octagon or of any other form provided it was uniform and that the Piston was of the same form and fitted it – it was not material whether the Condenser was on the right side or the left of the Steam vessel whether it was above it or below it provided it had the proper communication it was not material whether this Engine was set upon the one end or on the other – whether set in its natural position or turned upside down. . . .

'It was not material whether the communications were to be opened and shut by a Valve, a Cock a regulator of any denomination or any other mechanickal invention that could be applied to that Subject provided it was performed in the most convenient way that was known to Mechanics at the time – in short therefore no part of the construction of the Machine was at all material provided the principles of the invention were adhered to that there should be one Vessel filled with Steam rarefied to the degree of boiling water that that should Communicate with another Vessel in which it should be alternately condensed with every stroke to be made of the Pump and provided the Piston was made Air-tight by a Substance that would not evaporate as Water did provided these things were preserved every Mechanick in the Kingdom may construct this Machine in a thousand different shapes – in every different form that invention and imagination can suggest if he adheres to these principles still the Engine will have its effects and will be Mr. Watt's Engine.'

[Mr. Serjeant Le Blanc, counsel for the defendant, attacked Boulton and Watt's patent and specification and Parliamentary monopoly on several grounds.]

'Gentlemen [of the jury] – The question you will have to determine will be. – First whether this Invention as it is called or this Machine for which Mr. Watt has obtained a Patent was in itself perfectly new at the time he obtained the Patent. Secondly, whether he had done that which every Person ought to do before he can entitle himself to this favor from the Crown to be able to sell to the Public his Invention at his own price by Communicating a fair and full Specification so as to enable them without the Assistance of any of his Works to have the advantage of his Invention when his Patent should be expired. The third Object of [t]his Inquiry will be whether the present Defendant has in what he has done Infringed upon this Invention which is stated to belong to Mr. Watt and which by a Subsequent Assignment from Mr. Watt to Mr. Boulton he is become entitled to a part of. . . .

'Where a person applies for a Patent for the privilege of an Exclusive Enjoyment for a limited time of what he States to be his Invention it is necessary that the whole of that which he states to be an Invention shall be new in every respect. . . .

'. . . Now every one of the Witnesses which have been called on the part of the plaintiffs seem to have considered this [invention] as entirely new in every particular which is stated . . . [In regard to the air pump] they dont pretend to say that the Pump itself was a new Invention but the application of a Pump for the purpose of drawing out the Air or other Elastic [vapours] not so Condensed by the Cold of the Condenser. I am surprised that these Gentlemen who have stated that they have employed their Lives in

Studying the Steam Engine and in reading upon Mechanical Subjects have not found out that this very contrivance was the Subject of another Patent and that the very description of it as near as may be in the words of Mr. Watt himself now remains in the Court of Chancery upon a Patent obtained some Years before Mr. Watt applied for the present Patent.

'In the year 1759 . . . the Reverend Mr. Wood of Shropshire obtained a Patent for the Invention of a Fire Engine working upon a new Principle and at less than half the Expence of Coals hitherto used. – This Specification . . . states that the Condensed Air must be pumped out by a Pump to be worked by the Motion of the Engine or otherwise [and] Mr. Watt appears to have taken the very words of Mr. Wood when he states that the Air or Vapour not Condensed by the Cold of the Condenser is to be drawn out of the Steam Vessels or Condensers by means of Pumps wrought by the Engines themselves or otherwise . . . here then is a material part of this Machine which Mr. Watt pledged himself to the public he was the Sole Inventor of which was Invented by another Man and a Patent obtained for it long before Mr. Watt applied for his Patent.

'The next Article upon which some Stress was laid was that [in Newcomen's and other engines] . . . the Cylinder was open at the top and that till Mr. Watts Improvements it had never entered into the Idea of anyone to prevent the Pressure of the Atmosphere upon the Upper Surface of the Piston and to surround the Piston with an entire enclosed Vessel. I asked several of the Gentlemen whether in their Reading or in their Experience they had not heard of such an Invention but it appears it had escaped their knowledge and they consider that to be a material part of the Improvement and Invention of Mr. Watt that the Steam Vessel in which the Piston works was to be covered at the top so as to prevent the pressure of the External Air and merely to leave the Piston to be worked up and down by the pressure of the Steam at top or under it.

'In Newcomens Engine it is clear that the top of the Cylinder was open. . . . But in a Dictionary of Arts and Sciences which is published long before that there is a plate with a particular description of an Engine upon a small Scale certainly not applied for raising water out of Mines to the depths for which these Engines are applied but in the very description in the Universal Dictionary of Arts and Sciences there is a Plate of a Piston working in an Enclosed Steam Vessel or enclosed Cylinder where the description in the Specification if I may so call it expressly shows that it was to be worked by the force of the Steam excluding the operation of the Atmosphere upon this Piston. And that the top of the Vessel in which the Piston worked was closed that there was a clack at the top so that at the first Stroke to set it agoing the Air naturally was thrown out of it and then it worked by the mere pressure of the Steam without any Effect from the Outward Air. This Book . . . will be clearly proved to you to have been published . . . long before Mr. Watt obtained his Patent. Here is another part then of that which they state to be a new Invention which was clearly known and in the hands and Libraries of many Men of Letters long before this Patent was applied for . . . it will appear perfectly clear when this Book shall be read to you and you see the figure in the Plate annexed to it, that it expressly describes an Engine working with the Pressure of the Steam and not the

Atmosphere upon it and in addition to that the same description will be found in the Philosophia Britanica from which it is originally taken and another published in 1747 and in Harriss Lexicon Technicum the same description is given . . .

'The last Article in which they contend that we have infringed their Patent is the application of Oils Wax and fat Substances to aid the working of the piston . . . for the purpose of preventing any Air or Steam escaping from one part of the Cylinder to the other. . . .

'Now I shall have no difficulty in Showing to you that in the Engines which were in use long before Mr. Watts Invention many Persons applied . . . Tallow and Grease instead of Water for the purpose of making it work better and I will call to you Persons who have made use of Tallow and other Grease . . . without using any water this therefore is no new Invention introduced for the first time by Mr. Watt. . . .

'We will next enquire, whether from this Specification Mr. Watt has conveyed to the public that Information which will be sufficient to enable them at all times to Enjoy the Benefit of his Invention without the trouble . . . of making repeated Experiments and . . . the expence of trying whether one way or another will do better but they are to have a faithfull discovery made in such a way that from that discovery remaining on paper which may be described in words and which may be explained in figures annexed to the Specification they may see clearly from those words and Figures what the Machine is that is to be made and that they may immediately make it without the Inconvenience of repeated Experiments and trouble before it can be brought to perfection.

'The Gentlemen seem to have bestowed all their force in supporting two Articles of their Specification – they state simply that the Steam is to be condensed in Vessels distinct from the Steam Vessels or Cylinders altho' occasionally communicating with them. [Counsel went on to point out the complete obscurity as to this 'occasional communication'] . . . Why are the public to be put to the trouble of making Experiments . . . Why not describe it . . . that it [the condenser] must be separated by a Valve or Cock which occasionally opened and occasionally shut . . . with respect to this part therefore it is deceiving the public because it is not telling them at the time he made the Specification in what manner this is to be done that is was necessary to be made the Subject of future Experiments the conduct of Mr. Watt himself shews because in the Act which he obtained five years after the passing of his Patent [actually in 1775], he himself states in the Recital of that Act the Expence and labour he had been at in applying his Invention to the Engines, which he wanted to have the whole benefit of, the Money he had expended, and the time bestowed before he could bring his Invention into Execution . . . Why if that was the Case he ought to have annexed to this Specification some Drawing or Plan of the manner in which this Invention was to be carried into Execution, which he says had caused him so much time and expence. When he was to have a Monopoly granted to him for the Space of twenty five years, the public had a right to have that communication from him which had been the fruit of his Experiments and the fruit of his Expence . . .

'My Brother [Mr. Serjeant Adair, counsel for the plaintiffs] stated that the public could be at no loss because Messrs. Watt and Boulton had erected these Engines in different parts of the Country where they might at any time resort and see them, [but] what the Law requires is not that a Patentee should Erect in other parts of the Country to which by Possibility the public may have recourse or they may not but there is to remain in the Court of Chancery a Specification by consulting which alone any Mechanic of competent Ability may without making repeated Trials without repeated Experiments make that Engine for the Sale of which Mr. Watt has obtained a Monopoly for twenty five years, he [Watt] has not done this because he has not stated the particular Manner in which this Occasional Communication is to be made [nor had he provided technical descriptions or drawings of other parts of the engine]. . . .

'[Defendant's counsel went on to point out that although the eminent scientists such as De Luc, Herschel, and Lind, called as witnesses by the plaintiffs, had asserted that an improved engine could certainly be constructed from Watt's specification, they had been forced to admit] that the Specification was not sufficient to enable them to construct the Machine without repeated Experiments . . . I will call to you working Artists [artisans] ingenious Men who will tell you that from this Specification and without any information reading as it is here that they could not make this Machine which is to answer the purposes better than Newcomens Engine, that from trying a variety of Experiments they might at last perhaps attain to that which Mr. Watt has attained to of producing a Machine which would answer the purpose but that is not discharging the Duty which the Law imposes upon him – he is not merely to give a certain data upon which we are to work but he is to give some representation some figures by which we may without mispending Time and Money at once make that Machine . . . [Defendant's counsel also pointed out that part of the specification, that dealing with the 'steam wheel' or rotative engine, had apparently never been put into execution] and where a Man obtains a Patent for one joint Invention and its appears that any part of that is so described that an Artist cannot act upon it and make a Machine like it the whole Patent must fall to the ground. . . .

'[Defendant's counsel also argued that Bull's engine differed in certain material respects from Watt's engine, *e.g.* the cylinder was not enclosed in a steam jacket, it was inverted, and there was a different communication between the cylinder and condenser. But the jury's verdict eventually went in favour of the plaintiffs. In the words of Lord Chief Justice Eyre:] I think it is perfectly clear upon the whole of the Evidence that this is a very Noble improvement made by these Gentlemen in the Fire Engine for which they undoubtedly deserve the thanks of the public, and all sorts of protection as far as the Law will or can protect them . . .

'I think that there is no doubt but that in substance this Defendant has pirated this invention – because the subterfuge of having turned his Engine upside down instead of working it exactly in the Way in which it has been worked by these Gentlemen which upon the Evidence and upon the nature of the thing makes no sort of Alteration with respect to the operation of these new discovered principles or rather the application of old principles in a new manner does not at all depend upon it – it is used exactly in the same manner to all substantial purposes. . . .

'But I think there is a question behind and that question is admitting it to be clear that there is an invention admitting it to be clear that there has been a pirating of that invention – admitting it to be clear too that there has been Letters patent yet it is the Language of the Law that these Letters patent can be of no avail unless there is a true Specification of that invention . . .

'I confess I have myself very great doubt whether this Specification is sufficient . . . my doubt is whether the Instrument to be protected must not be distinctly organized and distinctly described with its organization . . . there is that question [of Law] behind which the verdict of a Jury will not touch . . . I think the Jury would have no great difficulty in collecting from the whole of this Evidence that a good Mechanic would be able to constitute an Engine from this Specification which would produce the Effect of lessening the Consumption of Steam and fuel in Fire Engines . . . the only question is whether the particular Instrument must not be organized and described as organized. . . . Then the Jury find for the Plaintiff subject to the opinion of the Court on the validity of the Specification.'

40. Extracts from 'Boulton and Watt *versus* Bull. Copy from Mr. Gurney's Short-hand Notes of the Argument in the Court of Common Pleas, June 27 1794.'
MS. Doldowlod.

This reserved point of law – as to the validity or otherwise of Watt's specification – came up for special decision by the Court of Common Pleas in June 1794. Boulton and Watt now began to change their stance somewhat, arguing not only that their specification laid down basic engineering principles, especially the principle of a separate condenser – thus rendering any variants of their engine, such as those of Bull and Hornblower, liable to prosecution as piracies – but also emphasizing more strongly that it provided sufficient technical directions for construction of improved engines, consuming less steam and fuel than Newcomen engines. Whereas their Counsel had argued in June 1793 that 'the most ignorant and stupid Person upon Earth', indeed 'every Man', could, the moment he was told of the separate condenser, immediately put this principle into effect and construct an improved engine, they now argued, as Dr. Small had argued at the time when the specification was drafted in 1769, that they were 'certainly not obliged to teach any block-head in the nation to construct masterly engines' (see above, pp. 55-6), but that their specification necessarily presumed knowledge of existing engines and engineering techniques, and that they were not therefore required to describe cylinder, valves, air pump, or even the condenser, in the details and arrangement of which there could be innumerable variants on the broad engineering principles laid down. Their specification had been drawn up for the direction of knowledgeable and skilful mechanics. Had Watt provided precise technical details, instead of the general principles, of his engine in the 1769 specification, he would have been wide open to innumerable piracies, by means of minor engineering variations, as Counsel for the Plaintiffs (Mr. Serjeant Watson) now pointed out:

'. . . the more precise the description of one form of an organized instrument might be – if that alone were to circumscribe the patentees exclusive right – the more certainly exposed to invasion must it be if others were at liberty to exercise his invention by any distinct means which their own ingenuity might invent they might do so in a hundred forms without deviating a single iota from the principle which he had discovered.'

Defending Counsel (Mr. Serjeant Le Blanc), on the other hand, argued against the validity of a specification which merely laid down general pinciples, instead of describing an 'organized' engine. Moreover, he pointed out material discrepancies between the 1769 specification and the 1775 Act:

'But the Action is brought for an infringement of the Plaintiffs right under this Act of Parliament and therefore we must consider what the Act is. It professes in the first place to be an Act for vesting in James Watt Engineer his Executors etc. the sole use and property of certain Steam Engines commonly called Fire Engines of his invention – professing therefore to be an Act for vesting in him the sole property in the whole instrument.

'It then recites that he had by Letters Patent in the ninth year of the King [1769] granted to him the sole benefit of making and vending certain Engines – not the sole benefit of making and vending an invention simply or making and vending a principle but it recites that he had granted to him the sole benefit of making certain Engines. It then states further that he had enrolled a particular description of the said Engine which description is as follows – stating the Specification – And then after reciting that he had been at a considerable expense in making complex machines and repeated Trials and that he could not complete his invention before the end of the year 1774 when he had finished some large Engines as specimens of his construction and whereas his Engines may be of great utility therefore the Act enacts this that the sole privilege and advantage of making constructing and selling the said Engines shall be vested in James Watt and his Executors for twenty five years it is therefore under this Act . . . that the present Action is to be maintained if at all. . . .

'What then is the privilege that is vested in him considering it under the Act of Parliament. Is it a privilege of applying a principle or is it a privilege of the sole making and vending an Engine . . . a regular Machine consisting of divers mechanical parts contributing to one purpose. It is the privilege of constructing and making and selling this [an Engine] which is granted to Mr. Watt. . . .

'The first ground therefore which I submit to the Court is upon this Act that he cannot claim a right to the sole privilege of vending the entire Machine when it is stated and admitted on his part that it is only an improvement or an addition to that Machine of which he is really the Inventor.

'Secondly considered upon this Act of Parliament alone whether there is any Specification which can support this grant and this privilege. . . . It is clear by the Act of Parliament that it is not open to the argument which was used yesterday that this is merely a principle and not an Engine or a formed Instrument . . . [Nor, in fact, was it possible under the Monopolies Act of 1624 to patent merely a principle, but only a new 'manufacture', which must be interpreted as a new method of actually making something. Watt's specification, however, was] not . . . descriptive of any Engine or Instrument whatever. . . . The reason seems obvious why this privilege of a monopoly which is to be granted by the Crown should not be granted merely for the Principle or for the first idea which may occur to an ingenious mind because if that is the case he is to reserve to himself the sole power of every possible improvement which may be made upon that idea in bringing it forward to perfection in the shape of a complete instrument – he is to have the benefit of it and yet at the expiration of that term the Public are not to have the benefit of that invention but they would be put to all the trouble and expence which he the patentee had been and for which he had been entitled to the benefit of fourteen years monopoly of that invention. [So Watt's patent and specification could not legally be maintained for a principle: it must be for a 'formed instrument', for an organized 'steam engine' as stated in the Act of 1775] . . . and the Specification does not afford to persons looking merely at the Specification information sufficient to enable them to make the thing at the expiration of the term which a Specification ought to enable them to make . . . there is no drawing no model no possible description of any Engine or Instrument.'

In his reply (copied into this same manuscript record) Mr. Serjeant now tried to shift away from the basis of principle, arguing indeed that there was *no* new philosophical principle in Watt's engine – 'the principle is the same that is that this elastic vapour [steam] created by fire and made more elastic by an additional quantity of heat is thrown into this Engine and is made to act upon it with more effect by the steam being condensed in a separate Vessel' – Watt had simply thought of a new arrangement of previously 'known parts of a known Machine', a new 'method' of utilizing steam power. But Plaintiffs' Counsel was driven to repeated inconsistencies in the use of words and arguments, and to concede the discrepancies between the wording of the 1769 specification and the 1775 Act. The case, indeed, had now become so confusing that it was necessary to have a second argument before the Court of Common Pleas in January–February 1795. Meanwhile, Counsel for Boulton and Watt looked more carefully into the history and formulation of patent specifications, in order to demonstrate that the 1769 specification was in conformity with earlier and current legal practice.

The same ground and the same legal arguments were gone over again in the Court of

Common Pleas in January–February 1795, but still without result. The situation was so complex that the judges, arguing the pros and cons in May 1795, were evenly divided in their opinions and so could not pass judgment.

41. 'General Information respecting the Specification of Patents collected by the Plaintiffs in the Case of Boulton versus Bull. 25 March 1795.'
MS. Birmingham Reference Library.

[First describes the way in which the Plaintiffs have collected information relating to patents and specifications.]

From all the observations they have been able to collect, it appears that *chemical and philosophical Inventions have always been and still continue to be specified in general terms,* without drawings or minute descriptions of the apparatus by which the processes are to be performed; and indeed, unless the Patentee assumes that which is not his Invention, and takes upon him to teach what is publicly known before, they cannot be worded otherwise.

Of the *mechanical* Inventions, in the *earlier* Specifications a general description only is given, and for the most part without drawings. In those subsequent to the year 1781, the description is much more particular and is generally, though not always, accompanied with drawings.

The great majority of Patents for Steam Engines come under the head of *mechanical improvements* and indeed it may be said, that the original invention of Captain Savary and Mr. Watt's method of lessening the consumption of steam and fuel are the only *new applications of philosophical principles.* Of many of these Patents prior to Mr. Watt's in 1769, no Specifications appear to have been ever enrolled; of those examined, the whole are described in general Terms without accurate dimensions or proportions and some are without drawings. The same remarks apply to such Specifications as intervened between Mr. Watt's first patent in 1769 and his second in 1781 for a *mechanical* application of the powers of the Steam Engine, which was followed up by a third in 1782 and by a fourth in 1784 all for *mechanical* improvements upon the Steam Engine. It is believed, that the first of these affords *the most early instance* of a regularly organized description and accurate delineation of mechanical improvements upon the steam Engine, and both the subsequent ones resemble it in those qualities. As no doubts had at that time been started respecting the validity of Mr. Watt's original Patent and no objections raised against the mode of its specification, there is no reason to suppose that he was less inclined at the time he drew it up to put the public in full possession of his invention, than when he delivered in these organized specifications and it is fair to conclude that the difference originated merely from the one being a *philosophical* and the others, *mechanical* Inventions.

42. 'A View of the Objections which have been at various times urged against Mr. Watt's Specifications.'
MS. Doldowlod.

The legal battle now widened, when the struggle against Hornblower's and other piracies was also brought into the law courts. The same arguments were produced by the same Counsel in the years 1796–99, and much the same evidence was provided, mostly by the same witnesses, on both sides. The manuscript volumes of legal proceedings piled higher and higher, but fortunately the conflicting arguments were very conveniently summarized by Watt in collaboration with his solicitors, though of course from a biased viewpoint.

Mr. Bull in his Answer in Chancery to the Bill filed against him by Messrs. Boulton and Watt sums up his Objections in the following manner. –

Admits that he makes use of Steam to press up the piston in his Engine which he conceives he has a right to do and that it is not an Invasion of the said James Watt's invention Steam having been so applied and such application particularly described in Harris's Lexicon printed in the Year 1700 and Owen's Works and several other Philosophical Works And the said James Watt in his said Specification also states that he condenses his Steam by the Cold of the Condenser by application of Water or other cold Bodies but does not add that he injects any Water into the Condenser to condense the Steam or if Water was injected by what method that Water was to be discharged from the Condenser after it has so condensed the Steam and the said Plaintiff James Watt therein also states that the Pump was for the purpose of drawing out the Air or elastic Vapour but omits to mention its use in drawing off the Water injected and therefore the Defendant conceived and does still conceive and submits that he hath a right to make use of a Pump when he sees occasion for the purpose of drawing off the Water after it has condensed the Steam as the Plaintiff James Watt's Specification is silent on the Subject and that he had a right to draw out the Air or other elastic Vapour by means of a Pump that use of a pump being made known in a Specification inrolled in this Honorable Court by the Reverend Mr. Wood of and in the County of Salop under a Patent obtained by him in or about the Year 1759 and described nearly in the very words used by the Plaintiff James Watt so that it is improbable for the plaintiff James Watt to assert that such Idea originated with him and as to the mode of applying it to a Fire Engine the said James Watt's Specification is totally silent.

Messrs. Jabez Hornblower and Maberley in their Answer in Chancery state their Objections as follows viz.

Say they have been informed and verily believe and expect to be able to prove that the said James Watt was not as the Plaintiffs pretend the discoverers of the principles upon which the Plaintiff's Engines are constructed but that the same were known and in actual use before the said pretended Discovery and Invention and before the said Letters Patent.

Say they apprehend and believe that the said Specification inrolled by the said James Watt is defective or is no Specification in many material respects for Defendants say that the said James Watt has in his said Specification stated that in Engines which are to be worked wholly and partially by the Condensation of the Steam the Steam is to be condensed in Vessels or Cylinders altho' occasionally communicating with them but hath not in his said Specification inserted or set forth any description whatever of such Vessels which then ought to have been as also the form and capacity thereof the same being a matter of great importance to the perfection of the plaintiff's said Steam Engine nor has he inserted or set forth how or when the Vessels in which the Steam is to be condensed are to communicate with the Steam Vessels or Cylinders which is likewise matter of great importance to the well going and good effect of the said Engine and which no Engineer or other person could understand without a particular description or instruction or many very expensive Experiments and Defendants also say that in the said James Watt's Specification he has set forth that whatever Air or other elastic Vapour is not condensed by the cold of the Condenser and may impede the working of the Engine is to be drawn out of the Steam Vessels or Condensers by means of pumps wrought by the Engines themselves or otherwise but the Plaintiff James Watt hath not inserted or set forth any description whatever of the said Pumps nor what proportion they are to have to any other of the principal parts of the Engine or how many pumps are necessary nor how they are to be wrought by the Engines themselves nor hath the said Plaintiff James Watt set forth in what Cases he intended to employ the expansive force of Steam to press on the pistons or whatever may be used instead of them which Defendants humbly insist he ought to have done inasmuch as the said Specification implies that there are Cases to which that mode of applying the Steam is better or more preferable to others. –

Defendant Hornblower says he verily believes it is not possible to apply Steam to a piston in the manner described or expressed in and by the said Specification. –

Defendants also say that although the Plaintiff James Watt has in his said Specification inserted that where Motions round an Axis are required he made the Steam Vessels in form of hollow Rings or Circular Channels with proper Inlets and Outlets for the Steam mounted on horizontal Axles like the wheels of a Water Mill within which were placed a Number of Valves yet the said James Watt hath not in his said Specification inserted or set forth any description of the said Inlets or Outlets or Valves or what the same are or consist of or how many of such Inlets or Outlets or Valves are necessary or where the same should be placed and altho' he has also inserted in his said Specification that in the Steam Vessels were placed Weights so fitted to them as intirely to fill up a part or proportion of their Channels yet rendered capable of moving freely in them by the means there said to be after mentioned or specified the Plaintiff James Watt has not in such Specification mentioned or specified those means and although the Plaintiff James Watt has inserted that when the Steam is admitted in those Engines between the aforesaid Weights and the Valves it acts equally on both yet as he has not set forth how or by what means the Steam is to be so admitted And the Plaintiff James Watt has in his said Specification inserted that he intended in some Cases to apply a degree of Cold not capable of reducing the Steam to Water but what such Cases are or when and in what manner the degree of Cold is to be applied the Plaintiff James Watt hath omitted to set forth And moreover Defendants say that the Plaintiff James Watt hath not in or to his said Specification inserted or subjoined any drawing Plan or Dimensions of the different parts the said Engines which Defendants humbly insist he ought to have done and by reason of the omission thereof and of the several other omissions aforesaid Defendants also insist it is impossible for any person to conceive a proper Idea or gain that knowledge of the said Engine which is meant and intended the Public should gain by or from the said Specification after the expiration of the said Term granted by the said Letters Patent and Act of Parliament And Defendants also insist the description of the Plaintiffs Engine is so imperfectly set forth in the said Specification that it is impossible for any Engineer or other person to make or construct Steam or Fire Engines from such Description in truth of which Defendants crave leave to refer to the words of the Act of Parliament set forth as follows –

'And whereas the said James Watt hath employed many Years and a considerable 'part of his Fortune in making Experiments upon Steam and Steam Engines commonly 'called Fire Engines with a View to improve those very useful Machines by which 'several very considerable Advantages over the Common Steam Engines are acquired 'but upon Account of the many difficulties which always arise in the execution of such 'large and complicated Machines and of the long time requisite to make the necessary 'Trials he could not compleat his intention before the end of the year 1774.'

Which words are in Defendants opinions sufficient to the conviction of any unbiassed person that the said Specification was so incompetent to instruct any other Engineer in the Erection of the Plaintiff's Engines that it could not instruct the said James Watt himself and that Therefore the said Specification is not a proper Specification according to the said Letters Patent.

Their several Witnesses state their respective Objections by their different Affidavits as follows viz. Jonathan Hornblower Engineer. His Objections are these:
1. The Specification not sufficient without Experiments.
2. That Directions or Instructions could have been given in the form of a Specification or by Figures and References.
3. That several of the Principles were known and in use before the Patent. – viz.

This he states as of his own knowledge	(a) The use and substitution of Steam in lieu of the Atmosphere as the impelling power to move the Piston.
	(b) The use of Grease and Oil on the Piston for the facilitating its Motion within the Steam Vessel.
This from Hearsay	(c) The use and practice of surrounding the Cylinders or Steam Vessels with heated Bodies.
	(d) The application and use of Pumps for extracting such Air and elastic Vapour as would impede the Engine.
	(e) The application of Water for the purpose of condensing the Steam in such manner as to effect that purpose without permitting the Water to enter the Steam Vessel.

Jethro Hornblower Working Engineer. His Objections are:

1. Experiments necessary.
2. Some Articles inserted in the Specification for rendering the Piston or other parts of the Engine's Air and Steam tight which would if used absolutely render the Engine of no effect.

Jesse Hornblower Engineer. This Person seems to think nothing of the objections stated by his Brothers except as to the necessity of *experiments* and that a Specification by Drawings etc was practicable which are the *only* objections *he* states by which he appears to negative the other Objections.

Arthur Woolf Engineer. Servant to Jabez Hornblower. This Witness states three objections:

1. Experiments necessary.
2. Specification by Drawings etc. practicable.
3. A Pump used for extracting Air and condensed Vapour before Mr. Watt obtained his Patent by Henry Wood in 1759.

Jabez Hornblower the Younger calls himself Engineer. This Witness is a Young Man the son of the Defendant Jabez Hornblower. He seems determined to make up for the deficiency of his Uncle Jesse. He states a Multitude of Objections –

1. That the Specification does not describe the Engine by the Principles without stating the manner of applying the Principles so as to enable any other person to form an organized Instrument thereof.
2. Experiments necessary.
3. That the Specification of the Motions round an Axis is more calculated to perplex than inform the Reader.

4. That the working of an Engine by the alternate expansion and contraction of Steam was unknown to Mr. Watt himself or purposely kept back till this day.
5. Rosin Quicksilver and melted Metals highly injurious.
6. These Articles not used by the Plaintiffs but Oil and Grease only.
7. That Oil and Grease were used about Fire or Steam Engines long before the Plaintiffs Patent.
8. That the most complicated Machines may be explained by Figures or Drawings with References.
9. That the Specification does not afford the smallest information or instruction respecting the nature of the Invention especially the manner in which the same is to be performed.

Thomas Strode, Engineer and pumpmaker. His Objections are:

1. Experiments necessary.
2. That a Specification might have been given by Figures and References.
3. The Pump used before by Henry Wood.
4. Some of the Articles for rendering the piston and other parts of the Engine's Air and Steam tight would be highly injurious or totally destructive of the Engine.

David Watson Engineer. His Objections are:
1. Principles only and not the manner of applying them set forth.
2. Experiments necessary.
3. The fifth Article impracticable.
4. Alternate Contraction and expansion of Steam unknown to Mr. Watt or withheld to this day.
5. Rosin Quicksilver and melted Metals improper and not used by the Plaintiffs who use Oil and Grease only.
6. Grease used about Steam Engines before Mr. Watt's Patent.
7. That the most complicated Machine may be described by Words and Drawings.
8. That the Specification does not contain the smallest information or Instruction etc. especially as to the manner in which the Invention is to be performed.

Thomas Rowntree, Engine Maker. A Witness on the first Trial. His Objections are:
1. Henry Wood's Pump used before Mr. Watt's Invention.
2. The expansive force of Steam adopted by Captain Savary in 1701.
3. The general manner of the Specification intended to take in any future Improvement.

Joseph Bramah. This eminent Water Closet Maker who calls himself an Engineer states his Objections as follows:
1. Want of Organization.
2. The Rotary Steam Wheel could not possibly answer the purposes of a Machine.

186

3. Want of Organization *repeated.*
4. That a Specification by Figures and References was practicable.
5. The Prior use of some parts.

Note. { (a) The pump by Henry Wood.
(b) The Steams acting instead of the Atmosphere was adopted by Captain Savary.

6. Melted Metals etc. improper.

Mr. Bramah winds up his Objections with a long Paragraph of which it is impossible to give any Idea whatever but in his own words which are much too curious to be passed over.

Saith that the ultimate effect of all Mechanical Apparatus and most especially of the Steam Engine depends on peculiar Organization and Preparation and on this Foundation rests all Invention and unless those two fundamental points be specifically ascertained by Delineation and Demonstration no Invention can be discriminately understood for this Deponent is of Opinion that to claim an Invention not described it may be any thing the Claimants find will best suit him And Deponent further thinks that the Plaintiff claiming his intended new applications of Steam and not describing his Engine is as though he had taken a Patent for every new application or combination of the Mechanic Powers without exception as every new effect in Mechanics is the effect of a different proportion and combination of parts possessing unalterable properties and on which some known force is to act so every new effect producible by the power of Steam or other Elements must bonâ fide depend on the peculiar proportion and organization of the Engine or Machine in which the said Steam etc. is to be employed as no new properties can be given to the Elements consequently all new effects producible by the varied application of their Powers must depend wholly on material construction and certainly such Construction can be rendered intelligible to Men of adequate Skill by Descriptions Drawings etc.

Two of these were Witnesses on the first Trial.

William Braithwaite
William Braithwaite
the Younger
and John Braithwaite. } Engine and Pump Makers. These Deponents jointly state the following objections viz.

1. Experiments necessary.
2. Specification by Figures and References practicable.

James Brown, Gentleman. This Gentleman's Objections are
1. Experiments necessary.
2. That the Specification does not state the application of Water to the inside of the Condenser nor how the Water was to be discharged.
3. That the Specification does not state in what cases the expansive Power of Steam was to be employed.

187

4. That he believes that the Plaintiff hath never been able to Manufacture an Engine moving on an Axis according to the fifth Article and that no one can from the description alone ever accomplish it.
5. Rosin and Quicksilver improper.

Messrs. Oxnam and others Co-Defendants with Mr. Bull in a suit in Chancery have summed up their Objections as follows viz.

1st. To the Invention. 2nd. To the Patent. 3rd. To the Specification. 4th. To the Act of Parliament.

Say they have heard that the said James Watt was not the original Inventor of the method of lessening the consumption of Steam and Fuel in the working of Steam or Fire Engines for which he applied for and obtained the said Letters Patent but that the same was Invented by Dr. Roebuck and that instead of their being intitled to such exclusive privilege that the said Patent was bad in Law for various reasons and amongst other Objections thereto Defendants are advised

That Patents cannot legally be granted for a principle merely as is the Case in the present instance or for a new improvement of Engines previously in use or if so that the improvement alledged to have been made by the said James Watt at least is not of the description which can be the subject of a patent even if discovered by him the same being a Principle merely and not a Manufacture and that his Majesty was therefore deceived in his Grant and particularly by not being informed that it was for an Improvement only and the said Letters patent are not within the protection of the Act of James 1st. and even if the said Patent was good which the Defendants do not admit that the said Complainant did not comply with the proviso therein contained by which it was required that the said Complainant James Watt should inroll a particular description of the said alledged Invention and the manner in which the same should be performed and used and that the Instrument inrolled by the said Complainant James Watt in the said Court of Chancery as aforesaid as such Specification is defective vague and uncertain and in many respects unintelligible and amongst the other Objections thereto that the said James Watt although he states therein 'That he condensed his Steam by the cold of the Condenser by the application of Water and other cold Bodies' yet does not add that he injects Water into the Condenser or by what means the Water was to be discharged from the Condenser after the Steam has been thereby condensed and although he therein states that the pump was for the purpose of drawing out the Air or elastic Vapours he omits to mention its further use in drawing off the Water injected into the Condenser and is silent as to the manner of applying the same to a Fire Engine And if the Complainants pretend that the said Letters Patent were granted not for a Principle but for a Machine or Engine then these Defendants are advised and submit that the said Specification is further defective in as much as no Engine is therein described or any representation of the alledged Invention or Improvement when applied or in what the addition to the Machines previously in use consisted or *where* it differed therefrom but defectively describes principles or Methods only and those he had in contemplation rather than those he had actually adopted and that no Machines could have been made from such Specification with equal effect with the Machines made by the said Complainants without various experiments And these Defendants are further advised and submit that although the said Letters Patent were for a principle only the said Act of Parliament was obtained for a Fire Engine and therefore that the same was not contained thereby and that the Parliament was deceived therein And Defendants further say they have been informed and believe that by the said Act of Parliament it is amongst other things noted that the said Complainant could not compleat his Intention before the end of the Year 1774 when he finished some large Engines as Specimens of the Construction And that by the said Act it is provided that every objection in Law competent against the said Patent should be competent against the said Act to all intents and purposes except so far as related to the Term thereby granted And these Defendants are therefore advised and submit that in as much as the said James Watt did not after passing the said Act ever inroll any new Specification of the said alledged Improvements which are therein recited not to have been compleated till the end of the Year 1774 that the want of the Inrollment of such new Specification is an Objection competent to the said Act as the want of Specification or the Inrollment of a defective Specification is an objection in Law competent to the said Letters Patent and all Liberties and Advantages whatsoever thereby and by the said Act of Parliament granted have utterly ceased determined and become void and so remain.

Observations

Jonathan Hornblower who objects to Mr. Watt's Specification for want of Drawings etc. obtained a patent in the Year 1781 for a pretended Invention of which he Inrolled the following Specification.

'To all to whom these presents shall come etc. Now know ye that in compliance with
'the said Proviso and in pursuance of the said Statute I the said Jonathan Hornblower
'Do hereby declare that my said Invention is described in manner and form following
'that is to say First I use two Vessels in which the Steam is to act and which in other
'Steam Engines are generally called Cylinders Secondly I employ the Steam after it
'has acted in the first Vessel to operate a second time in the other and by permitting
'it to expand itself which I do by connecting the Vessels together and forming proper
'Channels and Apertures whereby the Steam shall occasionally go in and out of the
'said Vessels Thirdly I condense the Steam by causing it to pass in Contact with
'Metalline Surfaces while Water is applied to the opposite side Fourthly to discharge
'the Engine of the Water used to condense the Steam I suspend a Column of Water
'in a Tube or Vessel constructed for that purpose on the Principles of the Barometer
'the upper end having open communication with the Steam Vessels and the lower end
'being immersed into a Vessel or Water Fifthly to discharge the Air which enters the
'Steam Vessels with condensing Water or otherwise I introduce it into a separate
'Vessel whence it is protruded by the admission of Steam Sixthly That the condensed
'Vapour shall not remain in the Steam Vessel in which the Steam is Condensed I

'collect it into another Vessel which has open communication with the Steam Vessels
'and the Water in the Mine Reservoir or River Lastly in cases where the Atmosphere
'is to be employed to act on the Piston I use a Piston so constructed as to admit
'Steam round its periphery and in contact with the sides of a Steam Vessel thereby to
'prevent the external Air from passing in between the Piston and the sides of the
'Steam Vessel In witness whereof etc.'

Though several of the Witnesses now brought forward call themselves Engineers there
is not among them one Engineer of Character or ability. Most of those who call themselves
Engineers are mere Pump Makers and inferior Artists.

Whatever the Witness *Watson* knows of Engines he learnt from Boulton and Watt.
He is not a Man of Character and comparatively little knowledge.

Rowntree was a Witness on the first Trial.

Being asked by Mr. Rous whether he understood the effect of Steam he answered 'In
some Measure. *I do not call myself an Engineer.*' And yet Mr. Rous drew from this Man
such further replies as shewed that all the Mechanical means that were necessary to Mr.
Watt's Engine were in use in Newcomen's Engine and perfectly familiar to Mechanics.

Rowntree has been publicly accused by *Bramah* of *pirating* some of his *pretended*
Inventions.

Bramah. From this Man's Examination on the Trial it evidently appeared he came
prepared to make the most of the Objections yet even he admitted it was *a good Text to
enlarge upon.* He said 'I have no doubt from that Text I should have been able to improve
Newcomen's Engine.' 'The hints there given would have *much* forwarded my Improvements.' Being asked upon cross Examination whether a Mechanic commonly skilful could
make an Engine from the Specification, He said 'He might make an Engine.' And further
'I have no doubt but the hints here point out the defects *tacitly* of the Old Engine.' And
further. 'If I had had this put into my hand when it was first wrote I could have made a
good use of it probably I might have invented a much better Engine than is now known
in the World in consequence of it.'

All this is mixed with Observations of the Witness which are in fact *contradictory* to
the passages here quoted. The Witness having admitted that an Engineer could make an
Engine from the Specification though he could not tell what sort of an Engine he would
be extremely puzzled to shew how such an Artist could miss making an Engine *possessing
Mr. Watt's Improvements.*

The Braithwaites. John Braithwaite admitted upon the former Trial that he had never
made a Steam Engine.

His Brother and Partner *William.* Being asked whether from Mr. Watt's Specification
he could make the Engine that Messrs. Boulton and Watt use – He said '*I am not certain
that I could. I could make an Engine but I am not certain that it would come near that –*
'In making this Engine I should have a variety of experiments to make before I could
'bring it to any thing before I could know whether it would answer the purpose it was
'intended for.'

William Braithwaite upon his Cross Examination being asked 'Whether knowing the principle of Newcomen's Engine and seeing Mr. Watt's Specification could you make one to condense the Steam in a separate Vessel.'

He replied. 'Yes I think I could.'

Q. 'Could you draw out the Air by means of a Pump?'

A. 'This might be applied but not very likely in the way that he has pointed out.'

Q. 'Then I am to understand you that all which you could not do would be this – That you could not certainly apply it in that precise form in which you see it upon that Drawing?'

A. 'I could not.'

Q. 'But could you with Certainty produce the effect of pumping it out?'

A. 'Yes, I suppose I could.'

James Brown. Gentleman – This Witness not professing to be a practical Engineer is not likely to gain much Credit with a Jury when he states the Specification to be *insufficient* in *opposition* to practical Engineers of Eminence who say the Specification is sufficient.

43. 'Answers by Mr. Watt to the Objections made to his Specification.'

MS. Doldowlod.

1st By Bull

The Engine described in Harris's Lexicon Technicum is Savary's who indeed, did employ the Elastic power of Steam, to press upon the Water to be raised, but his Engine had no piston, nor any thing which resembled one, – (see the Book) and this Misrepresentation Mr. Watt endeavoured to obviate, by a Memorandum annexed to the Specification, which says that the 4th Article should not be understood to extend to any Engine where the Water to be raised, enters the Steam Vessel itself, or any Vessel having an open communication with it.

As to Martin and Owen's Works, it is presumed a Dictionary under that Name, is meant, which was produced in the Court of Common pleas, which contains nothing relative thereto, except a very bad description and small print of a Newcomen's Engine very incorrectly copied from Clare upon Fluids, and seems an Engraving of a Model for Philosophic Experiments (which Books see) and every true Mechanic must at once see that the Engine so described in Owen could not act at all.

The Specification of the Reverend Mr. Wood does not relate to a Steam Engine, but to an Engine to be wrought by heated Air or by Air and Steam mixed, by Air being forced through boiling Water. It is believed that such Invention is impracticable and was never carried into effect by any Body; at least Mr. Watt was ignorant of it altogether, until a few days before the last Trial.

As a general Answer to the Objections of the defects of the Specification, the best way seems to give a short History of the Invention and Mr. Watt's then Ideas of the Application to the existing Steam Engines.

Mr. Watt found, that a well made Brass Model of Newcomen's Engine consumed Quantities of Steam and Fuel, out of all reasonable or direct proportion with larger Engines. He consulted Desaguliers natural Philosophy and Belidor's Achitecture [sic] Hydraulique, the only Books from which he could hope for Information; he found that both of them reasoned learnedly but by no means satisfactorily, and that Desaguliers had committed a very gross Arithmetical Error in calculating the Bulk of Steam from the Water evaporated in a common Steam Engine which being rectified, it appeared next, that his Data, or assumed Facts, were false.

By a simple experiment Mr. Watt found what was the real Bulk of Water converted into Steam and from his Friend Dr. Black, he learnt what was the heat absorbed, and rendered latent by the conversion of Water into Steam, which the Doctor then publickly taught and had done for some Years. Experiments had been made long before by Doctor Cullen, Mr. John Robinson [Robison] and others in public Classes, which proved that Water when placed in an exhausted Receiver boiled, and was converted into Steam, at the heat of 70 or 80 Degrees of Fahrenheit's Thermometer, while it was well known that under the pressure of the Atmosphere, it required 212 degrees of Heat to make it Boil, and emit Steam capable of displacing the Air. – It was evident that under intermediate pressures, intermediate Degrees of Heat would be required to make it boil and that in the Steam Engine more or less Cold Water must be thrown in according to the Degree of Exhaustion which might be required, or in other Words according to the Number of Pounds per Inch the Engine was loaded to.

Newcomen's and Savary's Engines existed the latter were in general laid aside, on several Accounts but, the principal one seems to be, that the cold Water, the raising of which formed the Effect of the Engine, entered the Steam Vessel itself, which, in general, was not a Cylinder, but was of an Oval or Egg Form, and by cooling it destroyed a great Quantity of Steam, when it came next to be filled, which Desaguliersexpr essly notices. – This Engine, however, had an Injection of Cold Water to commence the Condensation of the Steam, and Savary seems to have been the Inventor of that valuable Article, but he also seems in some Cases to have condensed the Steam by pouring Cold Water on the outside of his Copper Steam Vessel.

In Newcomen's Engine, the Steam Vessel was a Cylinder or so meant to be: – A Piston was suspended, moveable in that Cylinder; this piston hung by Chains to the Arch of a strong double ended Lever, like a Scale Beam; to the other end of which, the Rods which wrought the Pumps were suspended in the like manner: – The Steam was admitted from a covered Boiler through a pipe, into the Cylinder below the Piston the Air was blown out by the Steam at a pipe near the bottom of the cylinder called the Snift: – The passage from the Boiler was shut: – Cold Water was spouted or injected into the Cylinder from a Cistern placed higher: The Steam was thus Condensed or rendered less elastic; the other end or Mouth of the Cylinder being open, the pressure of the atmosphere not being resisted by an equally elastic Fluid within or under the piston, weighed upon the latter and caused it to descend, which by means of the Lever drew up the Pump Rods and raised the Water: The Injection Cock or Valve was then shut, the Steam Regulator or Valve was opened, Steam was readmitted: the Equilibrium of Pressure upon the upper and under sides of the Piston was restored and the Superior Weight of the Pump Rods by means of the Great Lever or working Beam drew the Piston to the top of the Cylinder and the Operation recommenced.

When the Piston was at the bottom of the Cylinder the Air which entered with the Steam and with the Injection Water was blown out at the Snift and the hot Water left in the Cylinder was expelled through another Pipe called the Eduction Pipe which proceeded from the bottom of the Cylinder several Feet downwards and its lower end stood in a Cistern of Water and was furnished with a Valve to prevent Regress.

The Steam Valve and the Injection Cock were opened and shut by certain Mechanism called the Working Gear, which was put in Motion by means of Pegs in a piece of Wood which was hung to and moved with the working Beam and was called the Plug-Tree.

In order to supply the Engine with cold Water it wrought a Pump called a Jack Head Pump which was shut at the top by an Iron Cover and its pump Rod wrought through a Collar of Oakum which permitted the Rod to slide up and down while it precluded the Exit of the Water which was raised to a greater height through a Side Branch turned upwards.

Thus the latent heat of Steam was discovered and published by Dr. Black.

The Boiling of Water in vacuo at low degrees of Heat was discovered and published by Doctor Cullen, Mr. Robinson [Robison] and several others.

The elastic Powers of Steam were known to Hero of Alexandria and to many ancient Writers. The Steam Engine was invented by the Marquis of Worcester, Savary, Papin and Newcomen.

The means of confining Steam and the making Valves Cocks and Regulators were known to all of them. Pumps for drawing both Air and Water out of Vessels or Reservoirs were well known to every body. An Air Pump with a Piston Rod moving through a Collar was invented and published by Mr. Smeaton and the same method even before him was commonly used in the Jack Head Pumps of common Steam Engines and in other Machines.*

A Cylinder and moveable Piston was used in Newcomen's Engine. So were the Working Beam and Working Gear or Machinery for opening and shutting the Valves and Cocks.

The Steam was Condensed by a Spout of cold Water in Savary's and in Newcomen's Engines and as it is said in Desaguiliers cold Water was poured on the outside of the Steam Vessels for the same purpose. Every body knew that cold Bodies of all kinds Condensed Steam when they came in Contact with it.

There were Pipes in all these Engines which admitted the Steam from the Boiler and Cocks or Valves which Shut it out from that Vessel and in Newcomen's Engine there was a Pipe which conveyed away the Cold Water and a Valve which prevented its Regress.

* This relates to the Piston Rod of the Cylinder for in respect to the Air Pump it is not necessary though convenient that it be shut at Top.

The Diameters of these pipes which admitted the Steam and let out the Injection Water had been ascertained sufficiently near.

The size and form of Boilers which answered sufficiently well had been also ascertained.

Of all these things Mr. Watt must say 'non ea nostra voco' – The things that are his remain to be told.

He found by the Application of the Knowledge which has been mentioned, that the Cause of the great Consumption of Fuel was that the Cylinder being cooled by the injection Water that Vessel *must* condense a large Quantity of Steam when ever it was attempted to be again filled with Steam; That the Vacuum could not approach to Perfection without the Steam was cooled below One Hundred Degrees, and that such Cooling would increase the Evil complained of in a fourfold or greater Ratio because the Penetration of the Heat or Cold into the Cylinder would be as the Squares of the differences of the Heats between that Vessel and the Steam. How was this to be Avoided? He tried to make the Cylinders of Wood or other Materials which conduct Heat slowly but he could not prevent the Steam from coming into Contact with the comparatively cold Water which remain'd in the bottom of the Cylinder and which must be expelled by the Steam; besides his Wooden Cylinders did not seem likely to be of long Duration.

In such like Experiments he spent much time and more Money than was suitable to his Circumstances yet he made no Advances towards a beneficial Discovery. But the matter having got firm hold of his Mind and his Circumstances obliging him to make Exertions to regain what he had Spent, he turned the matter over in every Shape and laid it down as an Axiom, – *That to make a perfect Steam Engine it was necessary the Cylinder should be always as hot as the Steam which entered it, and that the Steam should be cooled down below 100 Degrees in order to exerts* [sic] *its full powers.* The Gain by such Construction would be double; first no Steam would be condensed on entering the Cylinder and Secondly the power exerted would be greater, as it was more cooled. The postulata, however, seemed to him to be incompatible, and he continued to grope in the Dark, misled by many an Ignis fatuus, 'till he considered that Steam being an elastic Fluid it must follow the Law of its kind; and that if there were two Vessels A and B of equal or other dimensions, one (A) filled with Steam and the other (B) exhausted, if a communication were opened between these Vessels, the Steam would rush from the full one into the empty one, and they, both, would remain half exhaus[t]ed (if the Vessels were equal) or be filled with Steam of half the Density: If then, into the Second Vessel (B) an injection of Cold Water was made, or cold Water applied to its outside in sufficient quantity, the portion of Steam which it contained, would be condensed or reduced to Water, and by the same Law of Nature that had operated before, more Steam would issue from A into B until the whole was condensed, and nearly a perfect Vacuum established in both Vessels, yet as the Cold Water had not entered or touched A, that Vessel would still retain its Heat.

This Idea once started the rest immediately occurred; The Vessel A being supposed to be the Cylinder, B would be the Vessel called now the Condenser. The Water Air etc. accumulated in B, he, immediately saw, could be discharged or drawn out by means of a pump, or the Water might be let run out by a pipe more than 34 feet long going downwards, and the Air might in that Case be expelled at a Valve by filling B with Water, provided the descending eduction pipe were shut mean while. On the whole, however, he approved the pump: Another difficulty appeared, which was the making the piston tight; that could not be done with Water as in Newcomen's Engines, for that might get in and evaporate and produce Steam; he therefore thought of Wax Oil and Similar Substances, as Substitutes, knowing that they would not evaporate in the heat of Boiling Water and for greater Security he proposed to employ the Steam itself, as the acting power on the Piston.

The Diameters of the pipes necessary to convey the Steam into and out of the Cylinder, he regulated from those in use, the Size of the Condenser he assumed at Random, as he did that of the Air pump, which it was evident must be larger than was necessary to contain the Water and probable quantity of Air.

All this passed in his Mind in the course of a few Hours, and in a few days he had a Model at Work with an inverted Cylinder which answered his expectations, and was, as far as he remembers, equal in its properties of saving Steam to any he has made since, though in point of Mechanism much inferior; very simple Cocks were employed as Regulators or Steam Valves, and his Air pump and Condenser were of Tin plate – His Cylinder, however, was good and of Brass, two Inches Diameter, and a foot long. The Cocks were turned by Hand instead of being wrought by the Engine.

If Mr. Watt is thought worthy of Credit, in this matter, and the facts are consequently allowed where was the mighty difficulty of putting the Invention into Execution, from still fewer Data, than he has set forth in his Specification? He is not so presumptuous as to think that there were not and are not Numbers of Mechanics in this Nation, who from the same or even fewer, Hints would have compleated a better Engine than he did. Mr. Bramah has proved he could, and Mr. Watt is inclined to believe him, but Mr. Watt does not pretend that any Body could have done it without thinking upon it, nor without much previous knowledge and some Experience of similar Things.

Had Mr. Watt been content with the Mechanism of Steam Engines as they then stood his Machine might soon have been brought before the public but his Mind run upon making Engines *cheap*, as well as *good* and he had a great hankering after inverted Cylinders and other Modifications of his Invention, which his want of Experience in the practice of Mechanics, *in great*, flattered him would prove more commodious than his matured Experience has shewn them to be; he tried therefore too many fruitless Experiments on such Variations. He wanted Experience in the construction of large Machines, that he endeavoured to acquire, but experimental Knowledge is slow Growth, and with all his Ingenuity so much boasted to his prejudice, he was concerned in making some very indifferent Common Engines. Other Avocations, to him necessary, obliged him to turn his Attention from the Subject, 'till he obtained the patent, so that at that time he had made no Advances in the Improvement of the Mechanism. He therefore thought it proper to specify only what was his Invention, and to leave any Mechanical Improvements he might make to be secured by other Patents if worthy of them.

His Idea then was to apply his Invention to the Steam Engines as they existed; for this purpose there was nothing else necessary than to shut up the Snift, to apply a Regulator or Valve to the opening of the Eduction pipe within the Cylinder an air Pump to the outer end of that Pipe, and to inject into the upper part of the Eduction pipe. If at the same time, the Cylinder was defended from the cold of the Atmosphere, the Engine would thus be compleat, if the Weight of the Atmosphere were to be employed as the acting power for all the Regulators could be easily opened and shut by the then existing Contrivances and the Air Pump Rod could be suspended from the working Beam.

If however the Engine was wanted to receive all the advantages of the Invention the Cylinder was to be placed in a Case containing Steam, with Access for that Fluid to the upper side of the Piston so that it might act upon it, as the Atmosphere acted in common Engines or in the Case just stated.

And in this latter manner were the Engines made which he constructed in the beginning of the business, that is to say The cylinders were fixed in a case containing Steam with which Fluid they were wholly surrounded and their Mouths being open within the Case, the Steam had always access to the upper side of the Piston, and was admitted to the part below the Piston, only when the Piston was rising. The opening from the Cylinder into the Eduction Pipe was shut by a Valve, while the Piston was rising but when it was required to descend, the Valve was opened; these valves were of the sliding kind used in Newcomen's Engines. The Injection was made into the Eduction Pipe, and the Air Pumps, which drew out the Water as well as the Air, were fixed to the bottom of the Eduction Pipe which had a Valve to prevent Regress as usual. There were sometimes one pump, and sometimes two or three, as Circumstances or the fancy of the Moment directed. The Working Beams and working Gear were made in the usual manner, or nearly so, and in cases where there were Boilers fixed for the Common Engine which was superseded, they were used without alteration.

These Engines then differed in nothing from the ancient ones except in the Application of Mr. Watt's principles as set forth in his Specification.

It was found that the external Cylinder or Steam Case was very expensive. The Method of covering the Cylinder itself, with a Lid or Cover (which had been used in some of the Models) and conveying the Steam to the lower end of the Cylinder by a Pipe was adopted, and a less expensive Method of applying the Envelope of Steam was used, other kinds of Regulators were invented, and the whole Mechanism of the Engine was gradually improved, and these Improvements have been progressive for these last 21 Years. Some of them Mr. Watt has secured by other Patents but many of the most essential he has left free, and by means of them Newcomen's Engines have been improved to his Loss.

It will now it is hoped appear to the Candid that Mr. Watt has not wilfully concealed his Invention by a false specification but has set forth the nature of the same and the means of performing it, – he has told what he had Invented and it could not be expected that he should have described Mechanism already known to all Practitioners or not then invested [invented?]. Mr. Watt's Invention is merely a Contrivance to prevent cooling the Cylinder and to make the Vacuum more perfect by condensing the steam in a Vessel distinct from the Cylinder itself: – This is the nature of the Invention. The means of keeping the cylinder warm; The Substitution of the powers of Steam for those of the Atmosphere; Of Grease

196

etc. in place of Water to keep the Piston tight; and the drawing out the Air etc. by means of Pumps, are merely Aids in performing the principal Object.

This ought to be kept in view in judging of the specification; also; that Mr. Watt supposed it to be addressed to Mechanics and Philosophers and not to the ignorant –

Objections *considered*

Bull

Objects that Mr. Watt says he condenses the Steam by the Cold of the Condenser by the Application of Water or other Cold Bodies, but does not add that he injects any Water into the Condenser to condense the steam. In fact, Mr. Watt directs that Vessels shall be distinct from the steam vessels themselves and that these condensing Vessels shall be always kept as cold as the Atmosphere by the Application of Water and other Cold Bodies.

This evidently refers to the Common practice of an Injection, but it is also Fact that the Steam may be condensed by the external Application of Water to the Condenser, tho' not so expeditiously, unless the Surface of the Condenser be much extended, as Mr. Watt has found by many Experiments, but if sufficient time be allowed the Steam will be condensed even by a small surface; if kept always cold by a Stream of Water, and some Advantages would attend this practice, for then, no other Water would be required to be extracted by the Pump, than that quantity of which the Steam was composed. – The Injection was well known to every Body, who knew the Steam Engine, and if you had told the most ignorant Tender of the old Machine to condense the Steam (had he understood the word) he would have opened the Injection Cock. Mr. Watt's Silence therefore implies that it was to be done by the common means, unless the Engineer himself knew of a better.

When the Pump is pointed out for the purpose of drawing out the Air and other elastic Vapours, it seems ridiculous to object that it is not mentioned that it can also draw off the Water, such being the more general use of a Pump, and obvious to every understanding.

Hornblower Senior

Objects that the Size and Method of applying the Condensers were not specified. The nature of these Vessels does not admit of any determinate proportion to the Cylinder, at least such proportion is still unknown to Mr. Watt. – The Condenser will, to every thinking Engineer, appear to require to be large enough to hold all the Injection Water with some space to contain the rarified Air, which has entered with the Water or with the Steam. – It was well known before Mr. Watt's Invention that the Air contained in Common Water is about the 30th part of it's Bulk; consequently space must be provided for that, and it's Rarefaction by the Degree of Vacuum required, and as to the Air entering with the Steam it is so minute a Quantity as not to deserve Notice: That which comes in by the Air Leaks depends upon the perfection of the Workmanship, and is consequently incalculable and there ought to be very little if the Engine is good.

The same reasoning applies to the Air pump.

To those who will take the Trouble of Calculation it will appear that an Oversize in these Vessels is not hurtful. They have been in practice, made of various Sizes all with good Effect, and in the very first Trials Mr. Watt succeeded merely by following the Dictates of Common Sense, he knew that no precision was necessary, and could not suppose it would be demanded.

As to the form of the Condenser it is perfectly immaterial: a Globe or a Cube will do; a Cylindric or Tubular Form has been in general adopted as more convenient to be cast by the Founder or made by the Coppersmith, and in some degree from a kind of Deference for the ancient forms of the parts of Steam Engines. The forms of Pumps are well known, and that their working part must be made of some figure of equal Dimensions from end to end, whether its Area be circular or polygonal, and it is to be remembered that any kind or Species of Pump answers the purpose; The forcing Pump, and even an Iron Bellows have been used by some of the Pirates – but the common lifting Pump has generally been used by Mr. Watt.

The manner of working a Pump from the working Beam of a Steam Engine, was surely a matter unnecessary to be explained by mere words, and it seems a disgrace to the Profession, that any person calling himself an Engineer should complain of such want of precision.

In respect to the means of forming a Communication between the Cylinders and Condensers, which Mr. Watt certainly thought sufficiently obvious; All the Communications with the Cylinder in Newcomens Engines were made through or near the bottom of the Cylinder; there the Steam and the Injection Water entered and there the Eduction pipes had their Origin, and there Mr. Watt made his Openings for admitting and drawing out the Steam, but had this Question not admitted of this answer, it admits of another, which is that it may be drawn out at the upper End of the Cylinder, if that End is shut and performs the Office of the bottom as in Mr. Watt's inverted Models, and in Bulls Engines. The Absurdity of making Openings in any part of the Cylinder, where the Piston Slides through must appear too glaring for even a Tyro in Mechanics to attempt it; as to the form of the Passage it may be square, round, or angular at pleasure, and it has already been said that a common Sliding Valve would answer as well to open and shut it as it did in the ancient Engine*

The Specification saying that in many Cases the elastic Force of steam was intended to be applied to act upon the Pistons in the same manner, as the pressure of the Atmosphere was then employed, meant that in *most* Cases that was to be done, and indeed, there could be no Argument against it [in] all cases except some additional Expences which according to Mr. Watt's *Scotch* Ideas appeared a greater Object than it does to an English Tradesman. This Maxim was not stated, positively, because it was not an absolute Condition, a sine

* In applying the Pump to the Condenser to draw out Air and it may be applied to every part, and if it is wanted to draw out the Water. Scarce any person except these ingenious Witnesses would have required to be told, that it must be applied to the place where the Water was to be found, Viz. at the bottom or lower part of the Condenser.

quâ non; very good Engines may be made without it, hence the word 'Intend' which was used from mistaking the nature of a Specification, and alluding to Engines to be constructed in future, for it is a fact that in the Models Mr. Watt made before he obtained the Patent, he used the said force of Steam generally, and seldom the pressure of the Atmosphere.

Hornblower asserts, contrary to his better Knowledge, that it is impossible to apply Steam to the Piston in the manner described: If his Ingenuity or Knowledge did not extend to the Confining the steam by a Cover on the Cylinder, or by placing the Cylinder in a Vessel full of Steam he might if he had been inclined to do the thing, have placed the Cylinder within the boiler, where Steam had been usually confined, and it would have answered, as it has been proved in some small Engines†

To finish the Objections which relate to the essential parts of the Specification which are in question in the present Case, I come to the last Article, viz. the Substitutes for Water to keep the Pistons and other parts of the Engines tight.

The first Substance Mr. Watt had Recourse to was *Wax*, which answers indisputably well, – and there lies no Objections to it, but Price. Oils also answer very well, though some of them suffer Chemical Decompositions, which Mr. Watt was not then aware of, but which would not preclude their use if better ones were wanting. The Fat of Animals is the Substance generally used at this time.

Tar which is a resinous Body is also occasionally used, and Common Rosin might be used as it melts in boiling Water, and has then a lubricating quality. Quicksilver as a fluid impermeable to Air, might be used, were it cheaper, and if all the parts with which it was brought into Contact were made of Iron, it would not injure them as every body knows who knows the nature of the Substance. It has also been applied to make Pistons of Pumps tight and to lessen Friction, by Sir Samuel Morland in the last Century (see his Quicksilver Pump in Desaguiliers and other Writers). A Mixture of Lead Tin and Bismuth was published by Sir Isaac Newton, or his Disciples wnich melts in boiling Water and was commonly known under the name of Newton's Metal long before Mr. Watt's Invention, this would also answer the purpose of preventing the Passage of Air and Steam, though it might be attended with some Inconvenience and Expence.

There is an observation to be made here, upon which the Objectors are silent; it is, that these Substances are said not only to be fit for the purpose of keeping the Pistons tight, but also other parts of the Engines, meaning moving parts, and some of these substances though less proper in the Pistons, may be so in other places. Mr. Watt might shelter himself under the *superflua non laedant*, were that Grace extendible to an Engineer, but the fact is this, he tried the Substances mentioned and found they do in a better or worse Degree, and he inserted them in the Specification, lest Pirates should use them and thereby defeat his Patent. It is agreed on all hands that Grease Tallow and oil answer the end, and having found that, it was needless to have recourse to other Expedients in common Practice. There was no Intent to deceive, and it cannot be supposed that the short and

---

† It seems strange however that persons accustomed to Steam and used to confine it in Boilers, Cylinders Steam Pipes etc. could be at any loss how to confine it above the Piston.
Such Ignorance must be affected and wilful.

furtive Experiments of an Inventor, not secured by Patent could point out all the possible Chemical Contingencies which might happen to such Substances in continued use.

As to the objections to the 5th Article of the Specification Mr. Watt asserts and hopes to prove, that the Variety of his Invention there described, is in the first place intelligible, and secondly that it has been made and used with Effect. It was found however that the present Mode of producing rotative motions from reciprocating Engines was less expensive, and in several respects preferable, therefore the other was abandoned.

As to the 6th Article; (concerning which such impudent assertions have been made). It has never been executed *in great*, yet it is practicable, though it may not be preferable. It is well known that Steam like Air admits of Expansion by heat, and assumes a great Bulk. If then Steam of 300 Degrees of Heat were to fill a Steam Vessel, and then were drawn into a Condenser of 212 degrees of Heat, or that of boiling Water it would not be Condensed into Water though it would become less bulky, consequently a certain degree of Exhaustion would be made in the Cylinder and the External steam would Act upon the Piston with a certain force. In the same if steam of 212 degrees were to be drawn into a Condenser of 180 degrees containing Steam of half the usual Density its Elastic force would be lessened but it would not be Condensed to Water. If however Mr. Watt has been deceived and this principle should be thought either impracticable or erroneous, it ought not to invalidate the remainder of the specification as it stands perfectly distinct from it, and is introduced only as a Modification, and as nobody has infringed in that way it is not now in question.

On the whole of the specification it is to be observed that Mr. Watt's Invention being applicable to an infinite variety of forms of steam Engines, had little or nothing to do, with the Mechanism thereof.

The forms and situations of the parts allowing of great Variety as is sufficiently proved by Mr. Watt's own Application of his Invention and by the forms of the pirated Engines (which in the ingenious Mr. Flint's Words resemble each other little more than an Egg does an Oyster) and Mr. Watt having at the time of the Patent contrived no better Mechanism than was then in Use in the common Engine he was advised to the kind of specification he made, which if faulty in point of Language or Expression, should be, in some Degree executed [*sic*] by his then Comparative Ignorance of the English Language and even of literary Composition He must However still insist that the Method by him Invented is intelligibly set forth without Disguise. If some Men do not understand it, it may be, that Nature has denied them parts or Attention, or that their Ideas have run in other directions. He is also pursuaded that it is sufficient to enable an Instructed Mechanic to execute the Invention; perhaps as one person of Eminence observed, in a superior manner to Mr. Watt himself. To say that a specification must be sufficient to enable such person to execute it well, without *thought* or Trial, is exacting an Impossibility. Mr. Watt cannot now write such a specification and had he wrote one more detailed he would have probably more confounded than instructed his Readers. The Multiplicity of Words and of forms frequently misleads the Understanding. There are Specifications in the Office (Stanhope's) so much detailed and so long, that to read them once, or oftener, even with attention by no means gives clear Ideas of the thing intended. Mr. Watt speaks for himself but he has, in such Cases, found it the easiest way to find out the scope of the Invention and then to consider the means by which it may be effected in his own Mind, before he studies the Author's particular description, and when he has so done, it frequently appears that the Author's means are not the best he might have employed. Other Men may possibly make the same Observations on some of the Specifications for Mechanical Improvements on which Mr. Watt exerted all his powers of Description and spared no Labour.

To say that no Experiment should be necessary, is asking too much from any description of a complicated Machine or nice chemical Process. *The thing is in itself impossible.* Who can teach by words or Drawings any Man to paint like Sir Joshua Reynolds? and there is a Fact, a certain undescribable Sense or Feeling necessary in Mechanics as well as in painting.

Mr. Watt has seen and examined a specification of Mr. Bramah's Patent Lock which he thinks he understands (Repertory September 1796 No. 28) though he would probably have understood it as well and sooner without the Calculations and other Superfluous Verbiage with which it is confounded and perplexed, yet he is free to say that he could not thereby make a Lock equal to those made by Mr. Bramah without some abortive Trials and Experiments, if he could at all with his own hands; nor does Mr. Watt think a professed Workman could.

Plenty of other Examples might be adduced. An Engineer must think, he must calculate, and he must *invent*, even in the common Machines and other Matters of his business, in every Case he has not already practiced*

(Engineer ab exercendo Ingenium).

One thing might have misled many which is the looking at Engines constructed now by Mr. Watt, and finding in the Specification, no description of the Mechanism thereof. That Mechanism so far as it differs from Newcomen's has been the fruits of the Inventive faculties of himself and other Ingenious Men for twenty years; At the time of obtaining the patent, it was not invented; it did not exist, but all the essential parts of Mr. Watt's Method did exist and are exhibited in the Specification. No Law can be so unjust as to punish him for not describing what was not invented nor intended to be protected by the patent.

What has been said will account, it is hoped, for there being no Drawings to the Specification, Drawings can only exhibit particular forms and Dispositions, to none of which Mr. Watt could tie himself down, without manifest detriment, in an Invention, the

* Mr. Watt has lately contrived an apparatus for procuring factitious Airs for Medicinal Use. He published a detailed description of the Apparatus, and of the processes accompanied with plates, and moreover he prevailed on Boulton and Watt to make the Apparatus for those who require it. So that the using the Apparatus was all the Apothecary had to attend to. Mr. Watt wrote *con amore* and now sufficiently aware of the Difficulty of conveying Information by Words and drawings he was at all the pains he could to explain even Minutia. Yet he has had the misfortune not to be understood by many, the strangest blunders have been committed, and he has been obliged to write again and again other Directions, and after all, he believes insufficient for those who do not think or have not the necessary previous knowledge.

applications of which were so general and admitting of so many varieties, he was informed, however erroneously, that Variations of form might elude the Specification and that it had been so determined in the Courts of Law; he is happy to find now that the most respectable Judges think otherwise.

The preamble of the Act of Parliament is brought in to prove that he had still Experiments and Improvements to make. Granted, but what were they? Not upon the Method described in the Specification, but upon the Mechanism of the Engines which was defective upon the making perfect Cylinders Air pumps Regulators Valves etc. all which still admit of Improvement and long may do*

He had a more arduous Task still, which was to train Workmen to execute the parts, which is not yet perfectly accomplished. Mr. Watt does not pretend to answer for the Inaccuracy in the Title and preamble of the Act of parliament, these were the Work of the Clerks or imposed upon him in the Committee with whom he could not dispute if he had been fully aware of the Extent of the Impropriety – †

The accusations of Plagiarism are next to be considered 1st. It is said the Invention was made by Dr. Roebuck meanly [meaning?] probably Dr. Roebuck of Borrowstones. The Invention was made before Mr. Watt knew this Gentleman and Mr. Watt wanting support in money matters to carry on the Invention applied to him for that purpose and agreed to allow him a certain share of the profits upon certain Conditions but Dr. Roebuck to Mr. Watt's knowledge or belief never invented any part of the said Method of Saving Steam and Fuel. At another period Dr. Roebuck's Affairs not permitting him to afford the necessary Supplies he assigned all his Claims upon the Invention to Mr. Boulton and this is the whole of his Agency in it. Dr. Roebuck's eldest Son signs as a Witness to the Specification.

Steam was never used to press upon a piston, as an acting power in any Steam Engine that Mr. Watt knows of before his Invention. Savary's and Papin's applications of it have been explained.

Grease or Oil were some time used and some times Soap to lubricate the pistons of Newcomen's Engines, but always in Conjunction with Water and never as Mr. Watt believes or has been credibly informed for the purpose of keeping the piston Air Tight.

---

* Nor does the Act recite it was upon his Invention, but 'upon these large and complex Machines'.

† The Inference which Hornblower draws from the trials is intirely erroneous – The object of the Specification Viz. lessening the Consumption of Steam and Fuel was fully obtained in the first Trial where his Ideas were practised: But the desire of presenting to the world the fire engine not only improved in its principles but also in the Mechanical Construction induced him to employ many years of thought labour and expence in perfecting the valves Gear and other parts of its Mechanism. These are the Trials mentioned in the petition to the House of Commons as Grounds for the Compensation asked of them. They were then thought very distinct and sufficient Claims to their favour, which was shewn by prolonging the patent of the Original Invention and surely therefore the improvements which were considered by that Honourable House as additional Titles to reward cannot with Justice be admitted as a plea for invalidating his prior Specification. The public has been benefited by these subsequent Improvements not only by their Application to the Engines made upon Mr. W's principles but also by their introduction into the Common Engines and various other Machines. The Ingenuity displayed in the construction of the Engine is a Merit intirely distinct from the original Object of the patent and it has accordingly been separately awarded.

The Cylinders never were surrounded by any thing that Mr. Watt knows of, until Joseph Heaths took a patent in 1768 for surrounding them with Cold Water (This Man was Coal Overseer to Dr. Roebuck and probably had by some Treachery seen some of Mr. Watt's Models and mistook the use of the Steam Case).

The Application of pumps to extract the Air was equally unknown, and in the Old Engines unnecessary. Mr. Wood's project on that head has been explained to relate to an Engine working by means of heated Air and by it's alternate Expansion and Contraction.

The Condensing the Steam in a separate Vessel there is not the least reason to believe ever existed before Mr. Watt's Invention.

To Mr. Bramah's learned Declaration Mr. Watt will only answer that he has never claimed any thing under his Patent but what is contained in his Specification, otherwise he might have claimed Mr. Bramah's Water Closet according to Mr. Bramah's own Rule.

Granting that none of these separate Articles were new still the combination must appear to be so. The Condensation in a distinct Vessel could be of no use without the Air Pump or some similar contrivance for were the Air and Water to be blown out of the Condenser by Steam in the Old Way, the Evil of destruction of Steam would be augmented instead of remedied.

Were the Air Pump applied to an Engine with Injection into the Cylinder, and Water upon the Piston the Engine could be no longer mended.

Were Steam to be used as the acting power on the Piston in the last Case the advantage would be small if any and the keeping the Cylinder hot would do harm because it would occasion a greater Quantity of Injection Water to be used.

Some small advantage might possibly be derived from using Grease on the Piston but as the Water which enters that way serves as so much Injection Water it seems that the latter would be increased if the former were diminished something however might be gained in point of Friction.

Cylinders Pumps and working Beams, Boilers etc. are all old Inventions, yet the use of them is not alledged to weaken the Specification.

Then supposing the Action upon the Piston by Steam the keeping the Cylinder hot, The Condensation in a separate Vessel the Air Pump and the application of Grease to the Piston to have been all of them *separately* used prior to Mr. Watt's Invention they could not fail of being useless or hurtful but if they had been separately useful that does not lessen the Merit of the Combination of any two or more of them nor detract from it as a new and useful Improvement. Had they ever been used in Combination such use would have manifested its eminent Utility and would not have been laid aside nor would any Man have been found of such singular Modesty as to be silent for 27 Years while the Merit and *Profits* of the Invention were assumed by another.

No such Invention has however ever existed nor were even the principles used separately except in the Brains of these very ingenious Gentlemen Jonathan Hornblower and his Associates on whom it seems incumbent to prove by Facts instead of Hearsays and Beliefs that these principles have been used not only *separately* but *conjointly* before the date of Mr. Watt's Patent and it would further seem necessary that they were used usefully and publicly, private or infractuous use cannot serve their purpose and perhaps to obtain belief Witnesses may be required whose hands are clean and who have not themselves been guilty of Piracy. Not one word of all these prior Inventions was heard of until about 1780 when Mr. Watt's Invention was found to be useful and Hornblower Junior had commenced his Piracy and then, as far as Mr. Watt remembers the Murmurs were confined to the Grease and Hornblower or most of his Family gave their Opinions very freely against the Engine even after they had seen it and had its Construction explained. Some even after it was at Work. Captain Trevithick even to this hour or at least very lately maintained that Newcomen's Engines made by John Bouge were better.

44. 'Points necessary to be known by a Steam Engineer.'
MS. used in the Boulton and Watt versus Hornblower and Maberley Case, c. 1796. Doldowlod.

An especially noteworthy feature of Boulton and Watt's arguments in this case, as of the later proceedings in the Bull case, was the emphasis now placed on the assertion that Watt had drawn up his specification for the guidance of scientific and skilled engineers, who, with their knowledge of philosophical principles and current engineering practice, would certainly be able to use its directions to construct engines of a type superior to Newcomen's.

1st. The Laws of Mechanics as a Science.
2nd. Their usual practical Applications to the Construction of Machinery including the subsidiary Contrivance of Catches and detents etc.
3rd. The Means of making and constructing the various Parts of Machines in Wood from Brass and other Materials.
4th. The Law of Hydraulics and Hydrostatics, by which the Pressure of Columns of Fluids as well as their statical Weights are to be estimated; also the Quantities of any given Fluid or Liquid – which will pass thro' any given Aperture or Pipe in a given Time. The Resistance to motion or Vis Inertiae of Matter whether solid or fluid, and in the latter Case, how it is likely to be affected by the Form of the Channel or Pipe which conveys it etc. etc.
5th. The Doctrine of Heat and Cold, the relative Quantities of Heat imbibed by different Bodies, in acquiring an equal Number of Degrees of sensible Heat. The conducting Powers of different Bodies in Respect to Heat and Cold. The Quantity of Fuel necessary to heat a given Quantity of Water to the boiling Point. The Quantity of Heat necessary to convert it wholly into Steam of the Density of the Atmosphere. The Quantity of cold Water

necessary for its Condensation under like Circumstances. The Degrees of Heat at which Water boils under different Pressures.

6th. The Bulk of various Liquids especially Water when converted into Steam under given Pressures.

N. This is easily ascertained, with sufficient Accuracy, by any Person having a common Engine by measuring how many Fills of the Cylinder of Steam, the Boiler could yield per Minute when no Injection Water was used and measuring the Quantity of Water evaporated in that Time.

7th. The best practical Method of constructing Furnaces and seating Boilers so as to receive the proper Quantity of Heat from the Fuel. The relative Powers of various sorts of Fuel in producing Heat. The Methods of constructing and making Boilers and of adjusting their Sizes to the Cylinders they are to supply with Steam.

45. Extracts from 'A letter to the Rt. Hon. Sir James Eyre, Lord Chief Justice of the Common Pleas; on the subject of the cause Boulton and Watt *versus* Hornblower and Maberley: for infringement on Mr. Watt's Patent for an Improvement on the Steam Engine. By Joseph Bramah. Engineer.'
Printed. 1797. Birmingham Reference Library.

On the opposing side, Joseph Bramah, the later famous engineer, not merely gave evidence in support of Hornblower, as he had previously done for Bull, but also printed a very long statement addressed to Lord Chief Justice Eyre, in which he argued strongly against the sufficiency of Watt's patent. (At the same time, it must be pointed out that Bramah was himself an interested party, a rival engineer who wished to break Boulton and Watt's monopoly, and that he had to admit that Watt's specification would certainly help him to make an improved engine, though not without repeated experiments.) This document was published in 1797, for although the jury's verdict went in favour of Boulton and Watt against Hornblower in the Court of Common Pleas in December 1796, the same disputed point of law regarding Watt's specification, as had bedevilled the Bull case, was again brought before the Court in the following year, when Bramah sought to express his views more strongly and clearly by this printed pamphlet.

'. . . In considering the part arranged FIRST in this Specification, I cannot observe that the words there used, create . . . any new idea respecting the construction, proportion, or office, of that part of an Engine, properly called the steam cylinder. Nor do they tend to impregnate the understanding of a person well skilled in the fabrication of common Engines, with any image of an improvement . . . The inquirer is left wholly uninformed whether the intended cylinder, or steam vessel is to be left open at top, and shut at bottom or shut at top, and open at bottom; or whether both its ends are to be alike shut. Nor is he directed in what manner the steam is to be admitted into the cylinder, or in what manner discharged: there being no mention how, and in what part of the cylinder, the necessary inlets and outlets are to be contrived . . . There is likewise no mention made of the form, and action of the Piston, or the method of connecting it with the external and working parts of the machine; or whether the expansive force of the steam is exerted on the upper or under side of the said Piston; or even whether there is a Piston employed at all . . .

'The latter part of the above article . . . appears to be equally vague, and useless.

'It is there directed to "inclose the cylinder in a case of wood, or other materials which transmit heat slowly"; but there is not a hint given in any degree adequate to instruct the practitioner in what mode this case is to be applied; whether it must be in contact with the external surface of the said cylinder, or placed at some given distance from it. Thus . . . it is left to every Engine Builder to invent, and determine for himself. [Nor is there any explanation as to how the cylinder was to be surrounded by steam or other heated bodies.]. . . .

'From the words which compose the Article SECONDLY, respecting Condensation, . . . there is no description of . . . how this operation . . . is to be performed by a new method . . . and also to shew what condensers are . . . whether they are Tubes, Boxes, Cisterns, or what else . . .

'Article THIRDLY, sets forth that Pumps are to be used for extracting the uncondensed Vapour; but as in all the foregoing instances, the least trace of their situation, dimensions, construction or action is not mentioned; every Engineer must discover the best he can . . .

'Article FOURTHLY, proposes the Inventor's *intention*, or what he *means to do* rather than what *he has* done; but in this, as in the preceding instances, he has taken care to be equally incomprehensible; having resolved nothing positively, given no relation, proposed what is totally impracticable . . .

'Article FIFTHLY. This clause is a complete jumble of incoherent, unconnected, absurd, and indigested ideas; so blended and coagulated with mystery, ambiguity, and impossibility in practice, that it is a disgrace to the writer, and would undoubtedly ruin any mechanic who might attempt to analyze it.

'Article SIXTHLY. Here likewise the inventor states his *Intentions*, and not his *Actions*. And behold! What does he (by way of misleading) but propose what every man of Chymical Science must reject? viz. To work Engines by the partial expansion and contraction of steam. When it is a point long since determined, that there is no intermediate heat, or progressive operation, in the act of water expanding into steam [or of steam being condensed into water] . . .'

[Lastly, Bramah poured the utmost scorn upon Watt's proposal to use oils, wax, and resinous bodies, in addition to animal fats, and especially on the proposed use of quicksilver or other molten metals, for keeping the engine air and steam tight; the latter proposal was particularly impracticable and could only seriously damage the engine.]

[He went on to comment on the evidence given by De Luc, Herschel, Robison, etc. on behalf of the plaintiffs, emphasising that they were] 'not practical Engineers or Engine-builders . . . I am well persuaded so much depends on practice in this Art, that none of them could without much extra trouble and expence . . . even complete an Engine on the long-known principle of Newcomen . . . [How then could they construct Watt's much more complex engine] by the help of a Specification only, which . . . does not contain a single hint relative to proportion or organization . . .

'I think it is evident that after the principle, as Mr. Watt calls it, of condensing in a separate vessel had occurred to him . . . so much of this immense saving, they pretend to have made in Steam and Fuel depended on the due proportion, construction, and nature of the materials in these Engines, that it cost them six years (according to their own statement) of extreme study, labour, and an expense so ruinous, as frequently to oblige the patentee to make a stand for reimbursements, before they could organize the Machine so as to accomplish the desired effect.

'It must follow then, that the effect did not absolutely depend on nothing but the principles of proportion, and organization of the Engine; and that it might and ought to have been fully described in words, by drawings and references in the usual way.

'Whether they could not, or would not describe this Engine, is then the question. Admitting the former to be the case, the Engine must be a non-descript, and may be any thing; admitting the latter, which must be the case, the Specification is answerable to their wishes; for it is so abstruse and abridged, so mysterious and undecisive, that no human understanding can possibly penetrate it. In fine, it wants all that can be couched under the term specific, and may be said rather to include every thing than specify any thing.'

[Bramah then proceeded to demonstrate in copious engineering detail how essential it was to have a precise knowledge of the parts of the engine in order to build it successfully, and how many difficult engineering problems Watt had avoided in his specification. It appeared that] 'Mr. Watt took his patent not for what he had invented, but for what he might invent in future. Thus says he, "I will lay an indeterminate foundation, which will enable me to lock up the brains and hands of every other inventive genius; and if any have the hardihood to stir in the great field of improvement, to make any saving in the expence of fuel . . . by any means whatever, I will have at them with the hammer of the Law . . . [In other words, the so-called specification was made deliberately vague and general, so as to secure a complete monopoly.]'

Despite the strength of these opposing arguments, however, Boulton and Watt finally carried the day when their specification came up, on a writ of error from the Court of Common Pleas, for judgment in the Court of King's Bench in 1798–9, when the judges found unanimously in their favour. They agreed that Watt was the true and original inventor of a 'new manufacture', within the meaning of the Monopolies Act of 1624, and that the specification of his invention, method, machine, engine, or whatever it might be called, was sufficient to enable other engineers or mechanics to make it, and that it was not for mere principles.

46. 'Memorandums relative to Boulton and Watt's Engines in Cornwall, 20 April 1799.'
MS. Doldowlod.

An important consideration in the minds of judges and juries during the legal proceedings of the 1790s was the indisputable fact – despite legal and technical arguments over the terminology of the original patent specification – that Watt's improvements in the steam engine had given an immense impetus to the country's industrial development, especially in areas such as Cornwall, where coals were dear. At the same time, however, those industrialists who benefited so obviously from Watt's inventions tried to evade payment of the contractual royalties, while others utilized pirate engines. In the following document, of 1799, Boulton and Watt drew up a financial summary of the benefits their engine had brought to the Cornish mines, the royalties they had received, the technical and legal costs they had incurred, the financial concessions they had made, the premiums still owing, and the full extent of the obligations that were really due to them.

1. The savings in fuel alone (not including very large Savings in Boilers and Materials) effected by Boulton and Watt's Engines in comparison of what would have been required by Common Engines to do the same work during 20 Years from 1778 to 1798, have amounted to	from £800,000 to £1,000,000
2. The Premiums received by Boulton and Watt from Cornwall during the above 20 Years, have in all amounted to about	£100,000
3. The Premiums in Arrear now claimed by them amount to about £42,000, of which they may not recover the whole	£42,000
4. The Voluntary Concessions made by Boulton and Watt to miners in distress during the above period have amounted to upwards of	£35,000
5. Many Years and a considerable fortune were expended in getting the Engines introduced and the annual charge of carrying on their business, with which no person was debited, cannot be stated *Upon the whole* to have amounted to less than	£30,000 to £40,000

6. The opposition set on foot and supported by leading Adventurers in the Cornish Mines, has occasioned Boulton and Watt a loss of the greatest portion of their time during Seven Years and has obliged them to spend at Law and otherwise upwards of   £10,000

7. The quantity of Copper raised from the principal deep Mines worked by Boulton and Watt's Engines, (which could not have been raised otherwise without a very great increase in the price of Copper in five years from 1785 to 1790, amounted to £719,777, which being taken as an average of the twenty Years would have produced in all   from £2,500,000 to £3,000,000

8. If the Original Agreement of paying 1/3 of the savings had been strictly enforced and fairly and fully acted upon Boulton and Watt, would instead of £100,000 have received upwards of   £300,000

By a *correct* statement made in the Year 1793, it appears that Boulton and Watt's Engines compared with Common ones doing the same Work, effected Savings in One Year of £38,605 reckoning Coals at 42/– per Wey; Boulton and Watt's third part of which amounted to £12,868 but in fact they only received £5133 :: 7 :: 0.

And though the price of Coals has been stated at only 42/– per Wey (that being the price settled between the Adventurers and Boulton and Watt) their real cost was 63/– per Wey and the Amount consequently £57,907. Boulton and Watt's third part of which should have been £19,302 instead of £5133 :: 7 :: 0.

47. 'To the Honorable the Commons of Great Britain in Parliament assembled. The humble Petition of James Watt late of the City of Glasgow and now of Heathfield near Birmingham in the County of Stafford Engineer.'
MS. 1799. Doldowlod.

The above financial statement [46] was part of the effort made by Boulton and Watt in 1799 – after the conclusion of the long legal battle against pirates and just before their monopoly under the 1775 Act was due to expire – to secure a further extension of that Parliamentary privilege. Their main arguments were, firstly, that the development of the steam engine had taken many years of costly experiments, and secondly, that owing also to piracies and prolonged legal contests they had not hitherto been able to reap their just financial rewards. Parliament, however, rejected Watt's petition.

Sheweth

That his present most gracious Majesty by his Letters Patent under the Great Seal of

Great Britain bearing date at Westminster the fifth day of January in the Ninth Year of his Reign Did grant unto your Petitioner the sole Benefit and Advantage of making and vending your Petitioners Invention therein described, viz. 'A Method of lessening the consumption of Steam and Fuel in Fire Engines' for the Term of 14 Years.

That by an Act of Parliament passed in the fifteenth year of the Reign of his said present Majesty intitled 'An Act for vesting in James Watt Engineer his Executors 'Administrators and Assigns the sole use and property of certain Steam Engines commonly 'called Fire Engines of his Invention described in the said Act throughout his Majestys 'Dominions for a limitted Time' After reciting therein the said Letters Patent and that your Petitioner did, in pursuance of a certain Proviso therein referred to and in the said Letters Patent contained, cause a particular description of the said Method (which in the said Act is called by the Name of an Engine) to be inrolled in the high Court of Chancery at a Time therein mentioned, which description is inserted verbatim in the said Act of Parliament: And after reciting further as therein is recited, It was enacted that from and after the passing the said Act the sole privilege and advantage of making constructing and selling the said Engines therein before particularly described (that is to say, in the said specification so inrolled as aforesaid, and meaning the said Method of lessening the consumption of Steam and Fuel in Fire Engines) within the Kingdom of Great Britain and his Majesty's Colonies and Plantations abroad should be and was thereby declared to be vested in your Petitioner his Executors Administrators and Assigns for and during the Term of 25 Years from thence next ensuing. –

That in Order to carry into Effect the Intentions of the said Act of Parliament your Petitioner not having sufficient Funds of his own entered into an Agreement with Matthew Boulton of Soho near Birmingham Manufacturer to whom he assigned two third parts of the Benefits which might be derived from the said Parliamentary Privilege in conjunction with whom your Petitioner proceeded to erect these improved Engines for draining of Mines and such other useful purposes as reciprocating Motions were adapted to which Engines fully answered the proposed end of saving Steam and Fuel and also possessed other Advantages over the Common Engines, the saving so effected being from two thirds to three fourths of the Fuel consumed by the Engines which were before in Use. –

That your Petitioner Observing that the Mechanism of Steam Engines in General admitted of many Improvements he applied himself to perfect the same, which he did, first, by Applying the reciprocating Motion of the Engine to the producing of rotative or rotatory Motions whereby the powers of the Engine were applied to the turning of Mills of various kinds, to the very great advantage and increase of the Manufactures of this Country for which his Majesty was pleased to grant to your Petitioner his Letters Patent dated the 25th. Day of October 1781 – And secondly, he invented various other important Improvements upon the Mechanism and general construction of the said Engines and the Machinery connected therewith, for which he obtained his Majesty's Letters patent respectively dated the 12th. of March 1782 and the 28th. Day of April 1784 And, Thirdly, your Petitioner at various Times has made and perfected many essential improvements upon the Mechanism and Construction of various Parts of the Engines and other Matters

210

connected therewith for which he has never solicited any Letters Patent or exclusive Privilege and which have been applied to and have much improved the common Engines to the great advantage of the Public.

That the various Mechanical improvements made by your Petitioner have been the means of producing great ameliorations in the Construction and Workmanship of Mill Work and Machinery of various kinds throughout the Nation.

That the Terms on which your Petitioner and the said Matthew Boulton offered the use of the said Invention of Saving Steam and Fuel to the Public, were the acceptance of one third of the Value of the actual savings of the Fuel made by the Use of the said Invention when compared with common Engines performing the same Work, and that for the Use of the other Improvements made on the said Engines they have made no additional charge whatever having contented themselves with the said one third of his savings in Fuel.

That in the Course of their Business your Petitioner and the said Matthew Boulton have instructed many Engineers and Workmen who are now dispersed over the Country to the great improvement of the Arts.

That in making Trials to prove to the Public the superiority of their Engines and in experiments for making the said Improvements your Petitioner and the said Matthew Boulton have expended great Sums of Money and that the Labor and Attention of your Petitioner have been almost exclusively bestowed upon these Objects for the greater and best part of his Life and much to the injury of his Health.

That some years of the Term granted by the said Act of Parliament elapsed before the prejudices of the Public could be overcome, and your Petitioners Invention brought into general Use, during which the expenditure of your Petitioner and the said Matthew Boulton in making Trials to convince the Public of its Utility and in training Workmen etc. were very great and nearly proved ruinous to your Petitioner and the said Matthew Boulton, and no sooner were these Obstacles removed than piracies of the Invention were made, and the same countenanced by those who wished to rob your Petitioner and the said Matthew Boulton of the Benefits arising from the said Invention. These Piracies they found Means for some Time to repel in some degree without having recourse to Law, but about the year 1792 the Combinations formed against them became so formidable that they were compelled to have recourse to more Active Measures, First by soliciting the interposition of this Honorable House against a Bill brought in by one Jonathan Hornblower to prolong a patent he had obtained for a plagiarism of your petitioners said invention which Bill was withdrawn in consequence of an adverse Vote of this Honorable House, Secondly, by two Actions of Law and several Suits in Chancery against other plagiarists of your Petitioners said Invention and persons who supported them, in which Contests and Litigations your Petitioner and the said Matthew Boulton have been unfortunately engaged for these last Seven Years and until the disputes and questions which were raised therein respecting the validity of the said first mentioned Letters Patent were terminated by the unanimous decision of the Court of Kings Bench in favor of your Petitioner and the said Matthew Boulton in Hilary Term of this present year upon a Writ of Error from the Court of Common Pleas in the last of the said Actions.

That the Application to this Honorable House and the said Suits in Law and Equity have been attended with an enormous expence, great loss of Time, and much anxiety of Mind to your Petitioner and the said Matthew Boulton who have thereby been prevented from pursuing their Business in the Profitable Manner they might otherwise have done, and great Sums of Money due to them from various persons in the County of Cornwall have been and still are withheld from them, the damages and Costs they have obtained in the Actions at Law have been only nominal and partial and the Intentions of the Legislature in favor of your Petitioner have been in great Measure frustrated.

That the sole causes of such protracted Litigations have been the uncertainty of the Law regarding Patents and some pretended inaccuracies (intirely verbal) in the said Patent and in the Act granting to your Petitioner the said parliamentary privilege.

That in the Course of these Suits the Originality of the Invention of your Petitioner has been fully recognized and the Competency of the specification as well as the great utility and merit of the Invention established by most respectable Witnesses and the Verdicts of two special Juries.

That notwithstanding the said Judgment in their favor, your Petitioner and the said Matthew Boulton are now involved in fresh perplexities by a Question or doubt started in one of the said Suits now depending in Chancery namely, Whether your Petitioner has not intirely or in some degree forfeited or surrendered (by some sort of legal implication quite contrary to your Petitioners actual implication) the said Parliamentary Privilege by accepting the said Patent granted by his Majesty in 1782 for certain Mechanical Improvements on Steam Engines which Improvements are of another Nature from and were additional to his former Invention, for which he obtained the said parliamentary Privilege, and which were also applicable to other Steam Engines and to other Uses, and which Question or doubt has been referred to the decision of a Court of Law.

That your Petitioner is apprehensive that in case of an adverse decision on the said last mentioned Question he and the said Matthew Boulton may not only be prevented from recovering large Sums of Money now unjustly with held from them but may be nearly ruined by endless Litigation and at least the remainder of their Lives spent in bitter anxiety.

That inasmuch as no adequate recompense can ever be obtained in the Ordinary Course of Justice for the Toil Anxiety and Expence which have been most wrongfully brought upon your Petitioner by these long and expensive Contests in which he has been involved in the defence of the said Parliamentary Privelege; And inasmuch also as your Petitioners said Inventions have been of very great utility to the Mines and Manufactures of this Nation and as the Bounty of the Legislature designed to be conferred upon your Petitioner for the said Services have been in a great Degree frustrated by combinations formed against him and the said Matthew Boulton, your Petitioner humbly presumes to throw himself upon the protection of the Legislature, in this which appears to him a very uncommon case and peculiarly deserving of redress, and especially considering that the Questions hitherto have been merely formal and that the only one now remaining your Petitioner is involved in merely through accident and by mistake, and that too, in consequence of having made Improvements, which your Petitioners very Enemies will acknowledge to be important and for which your Petitioner has obtained no reward beyond the said third part of the saving of Steam and Fuel which was the agreed recompence for the use of the said Original Invention.

Therefore your Petitioner most humbly prays this Honorable House to grant leave to bring in a Bill for extending the Term of the said Parliamentary Privilege for a reasonable Time and for revising the said Patents granted to your Petitioner in the Years 1782 and 1784 and extending the Terms thereof in like manner for removing doubts and difficulties concerning the same or for granting such other relief in the premises as to this Honorable House shall in its Wisdom seem meet.

And your Petitioner shall ever pray etc.

48. James Watt, 'Thoughts upon Patents, or exclusive Privileges for new Inventions.'
MS. Birmingham Reference Library.

Even before Boulton and Watt had resorted to the courts in defence of the steam-engine patent, Watt had foreseen that difficulties would ensue from the confused state of the law and from deficiencies in the wording of his own Act of 1775. In February 1783 Watt suggested to Boulton the need for a new general Bill to explain in what way a specification should be drawn to be valid, and calling upon all holders of patents to submit new specifications in accordance with the Judges' new requirements. Nor was Watt alone in his fears for his property. The rejection of Arkwright's patents for carding and spinning in 1781, and again in 1785, was one among a number of factors causing a committee of patent-holders to begin an agitation for the reform of the law, while, on the other side, the Manchester Commercial Committee had been pronouncing on 'the pernicious effects of patents' at least since 1774 and was largely responsible for the defeat of Arkwright's claims. Watt gave evidence for Arkwright in February 1785 when his cause came before Lord Loughborough and again in June 1785 when the case was before the House of Lords, and he clearly felt that in assisting Arkwright, whom he did not like as a person, he was defending his own interests.

Watt and Arkwright seem to have consulted one another in drawing up a manuscript, now in the Birmingham Reference Library, entitled 'Heads of a Bill to explain and amend the laws relative to Letters Patent and grants of privileges for new Inventions', if we can accept Sir Erich Roll's assertion that the minor emendations to this document are in Arkwright's hand.* The manuscript tries to give greater precision to the phrase 'any manner of new manufactures', in the Statute of Monopolies, so that it would cover 'every new and useful

* Erich Roll, *An Early Experiment in Industrial Organisation* (1939), p. 146 n.2. See also Appendix VI, pp. 284–286.

Philosophical Chemical or Mechanical art or Invention whatsoever', to allow specifications to stand so long as they were sufficient to enable an artist skilled in the branch to carry them out, and to ensure that a patent for a new invention should not be rejected because it included descriptions of processes already known. It is evident that this document is earlier than the one which we print here, which is a much more rounded, complete and finished exposition of Watt's views on patents and on the reform of the patent laws. The preparation of this second document seems to date from 1786, since on 19 March 1786 Watt wrote to Boulton:

'In relation to the patent paper, I have not at present the ability to shorten it, that kind of work demands a clear head and vigorous mind neither of which I am at present blest with. The paper as it is can be read through in about half an hour.'*

It was probably this same paper which was submitted to the Lord Chancellor, Lord Kenyon, in August 1790.

Many eighteenth-century inventors were less fortunate than Watt in the defence of their patents. Arkwright lost his for carding and spinning in 1785, and Argand, who had consulted closely with Boulton and Watt about the specification of the patent for his lamp in 1784, also lost his patent in March 1786; Cort (puddling and rolling iron) and Tennant (chlorine bleaching) were among others whose patents were similarly quashed in later years. Nor had the situation improved at all by 1829, when the witnesses to the Select Committee on Patents testified to the uncertainty of the law.

When the Legislature restricted the Kings prerogative right of granting the Royal letters patent vesting in individuals the sole right of using or vending certain Commodities, they permitted the exercise of that prerogative to continue in granting patents to the *authors of new inventions* for the term of 14 Years, with a declared intention to stimulate ingenious men to improve the mechanical and chemical arts. Because it was supposed that men of good sense, and of limited fortunes wou'd not throw away their time and their money in endeavouring to bring an art of invention to perfection, unless they had a prospect of being amply repaid by making greater profits than they cou'd do in the common course of their business –

Few projectors and it may be said, that few men of ingenuity make fortunes, or even can keep themselves on a footing with the tradesman who follows the common tracks, and who possesses no other merit than that of attending solely to his immediate interest without suffering himself to think seriously whether the article he manufactures might, or might not be Improved. The reason is plain, the man of ingenuity in order to succeed in the object he takes in hand, must seclude himself from Society, he must devote the whole powers of his mind to that one object, he must persevere in spite of the many fruitless experiments he makes, and he must apply money to the expences of these experiments, which strict Pru-

* J. Watt to M. Boulton, 19 March 1786, A.O.L.B.

dence would dedicate to other purposes. By seclusion from the world he becomes ignorant of its manners, and unable to grapple with the more artful tradesman, who has applied the powers of *his* mind, not to the improvement of the commodity he deals in, but to the means of buying cheap and selling dear, or to the still less laudable purpose of oppressing such ingenious workmen as their ill fate may have thrown into his power.

Both these species of men have their use in a trading state, neither of them could so well subsist without the other; but the ingenious artist more particularly claims the protection of the legislature, because he is least able to protect himself, in the common concerns of life. He should be considered as an Infant, who cannot guard his own Rights, and as he has purchased his inventions *with his time*, his *money* and *his ingenuity*, and often also at the expence of his health and peace of mind, is it not just, that the exclusive privilege of using them shou'd be secured to him in such manner as either to enable him to turn them to profit himself, or if he has not the necessary abilities to do so, to enable him to dispose of his privilege, or to associate himself with others who are more hackneyed in the ways of the World? In granting this privilege the state gives nothing; if the invention is not found an improvement, people will soon cease to use it, and the inventor will be punished for his presumption by the very means by which he hoped to acquire money; if on the other hand the inventor acquires a fortune by it, is not that a proof, that the public have found their advantage in it? For otherwise they wou'd not have used it; and is not the person who has by his ingenuity and industry put the public in possession of such an advantage justly intitled to the money he may acquire by it?

It is argued by the opponents of patents, that they Tend to cramp ingenuity, by circumscribing the artist to the use of the arts which prescription has made public property. Those who argue in this manner have too narrow notions of the powers of the human mind, and of the objects on which it can exert itself, and must suppose that our ancestors have left us very little to add to the improvements they made in the Arts. The contrary is evidently fact, the improvements which have been made within the last 50 Years *surpass all which ever have been done in an equal period of time*, and we are far from finding that we approach to the ne plus ultra of invention. If we did, there woud be no room for patents, and the objection woud of itself be done away. The fact is, that men of limited abilities, who can only go on in the path which some body else has pointed out may find themselves confined by the patent which prevents their interfering with their neighbors invention, but though a man of true genius might envy another the honour of discovery, he wou'd rejoice in his gains; and it may very probably be an advantage to the state that men of inventive faculties being prevented by the strict enforcements of the rights of patentees from meddling with the inventions of others shou'd be oblig'd to strike out new improvements on other subjects. The field is surely wide enough.

But it is said that if any improvement can be made upon a patentees invention, it is hard that the public shou'd be deprived of the use of it until the patent expires. If it be good policy to grant a man a patent for his invention *Justice* requires that he shoud be protected in the exercise of it, otherwise the patent is a vile imposition practised by the state upon the patentee – But the good policy is disputed, let us see what the public gives and what it receives. It gives the patentee a right to prevent others from using a thing which it is supposed did not exist prior to the patent, and that only for the term of 14 Years. The state then loses nothing; because it has only secured the patentee in the possession of a thing which in respect to the state was a non entity, and that for a term which though it makes a considerable portion of a Man's life, is a very small one of the duration of a kingdom: On the contrary the public gains by having a new Art added to its stock, or an old one improved, human labour is abridg'd or the value of its productions encreased, and this without any expence to the public – Such being the facts, is it not a species of injustice to argue that every blockhead shall have a right to avail himself of his more inventive neighbours ingenuity experience and industry without making him any compensation for it, as must happen if patents were not granted to protect the inventor? or that by making an alteration or even improvement on the first idea (which in all probability he wou'd never have thought of if he had not had the advantage of his predecessors experience and ingenuity), he shou'd have a right to use any part of what is fairly the patentees invention or covered by his general description of the principles thereof, although not particularly described? And as no man can within the limits of a specification enumerate every possible variation which may be made on any mechanical or Chemical principle, is it right to Judge of the Validity of a patent by the strictest sense of the words of a specification, and to condemn the patentee because he has omitted to describe the particular shape of some part of the machine, the materials of which it is made, its exact dimensions, or the weight of some ingredient of a chemical composition, while it is evident from the fact that it is so well known to the publick that another has been enabled to make it? The state does not always gain by having new inventions made public property, where the commodity manufactured is an article of fashion or fancy and not of necessity, Competition among manufacturers will frequently be injurious to the nation in general for it will be exerted to make the article cheaper, which is commonly accomplished by making it slighter, or in a worse manner; so that partly from the lowness of the price which renders it despicable, and from its badness which deprives it of the small pretentions to beauty or utility it might otherwise have, the commodity falls into disrepute or neglect and the advantages which might have resulted from the Manufacture are totally lost to the state; instances of this kind are very frequent in the Birmingham and Manchester manufactures.

In other cases where an invention consists of a complicated machine or a difficult process which demands much experience expence and attention to bring it to perfection it is more likely to have these bestowed upon it by persons whose fortunes depend on its success, than on the publick in general who are disposed to go on in the beaten tract, in which they are sure to succeed to a certain degree and are not disposed to make Experiments uncertain in their event and for which they never can be repaid by any advantages in the use of the machine or process so far as they are concerned individually –

The Validity of patents is at present commonly understood to depend principally on 4 circumstances 1st. The novelty of the invention 2nd. The patentee being the true inventor 3d. The same not being practised by others at the time of granting the patent 4th. The nature of the invention and the manner of performing it being clearly specified by a writing or drawing given into Chancery. But though these circumstances are allowed to be necessary to the validity of the patent, yet they are so ill defined by either law or custom that a patentee never knows when he is secure, particularly in the article of the specification, on which many of the late decisions have turned –

Let us therefore consider how the Law now stands and how in justice to the inventors, and to the public it ought to stand.

Previous to the reign of James the 1st. the Kings of England were in use to grant Patents, not only for the sole use of new Inventions, but also for the *sole making and vending of many Commodities which had been from time immemorial in common use*, but as the exercise of the latter part of the prerogative had produced many vexatious and grievous monopolies, an Act of Parliament was made in the 21 of that Kings reign, taking away that branch of the Kings prerogative which related to the granting monopolies of Articles in *common use*, but preserving and confirming it in respect to 'Letters patent and grants of privilege for 'the term of 14 Years or under for the sole working or making of any new manufactures 'within this realm, to the first and true inventors of such manufactures *which others at the 'time of granting such letters patent shall not use*, so as they be not contrary to Law nor 'mischeivous to the State by raising prices of commodities at home nor hurt of trade, or 'generally inconvenient, the said 14 Years to be accounted from the date of such letters 'patent, or grant of privilege, but that the same *shall be of such force and effect as if this act 'had never been made; and of no other*' –

Since that period, it does not appear that any other Statute has been made, farther circumscribing the Kings prerogative, and upon that, patents now rest, as appears from the last words of the clause – But it is worthy of remark that here we find no mention of specifications nor has any Statute been pointed out, by which they are required, and it does not appear by searching into the records of Chancery, that any specifications are enrolled there before the year 1714. It appears however that the ancient patents had their titles more fully worded so as to comprehend what might be called the *Principles* or Grounds on which the invention proceeded, and in some cases the inventors were required to publish books, when the inventions was intricate and difficult to be distinguished from those in common use, but in all cases the enforcing such disclosure was the Act of the King. If therefore as is apprehended there is no statute law enforcing such disclosure, the necessity rests solely on the proviso in the patent, which requires it, and in the opinion of the writer, which however he gives with due deference to the Gentlemen learned in the Law, it seems to have been originally intended not so much as to secure the public in the secret of the invention, *as to discriminate one inventors property from that of another*, and from that which belonged to the Publick by being in common use, and also to prevent a second patent from being valid for the same invention. Be the original intention of the Clause what it will, the words in the patents which require the specification are as follows, and it is presumed, that it is to these, and these only that the inventor is to attend, for he has no other direction given him on the part of the King, and it is not to be supposed that every inventor is Lawyer enough to know what sense the Courts of justice have put upon the clause, particularly, as few reports of their decisions stand upon record in the Law books.

'Provided also that the said A.B. shall particularly describe and *ascertain the nature* of 'his said invention and in what manner the same is to be performed by an instrument in writing' – It seems by these words, that the inventor is required to ascertain the *nature* that is the new *principles* on which he proceeds, and the manner in which he uses these principles, so that others may not offend by making use of them unknowingly, but it is not said that he shall describe them in such a manner that *any other person* shall be enabled by the *specifications alone* to make the machine or perform the process, without any previous knowledge of the subject. If such were the Law, it wou'd be next to impossible to describe any complicated machine, or any nice chemical process so that the specifications shou'd not be set aside for want of being sufficiently minute or clear; and if any inventor cou'd attain to such exactness and clearness of description and not be allowed to secure himself by a general and comprehensive description of the principles, might not any alteration in forms or proportions be deemed a legal evasion of the patent? Equity seems therefore to require, that the new principles shou'd be clearly ascertained, that the invention may be discriminated from those of others, and that the manner of using them shou'd either be obvious or so described that persons of Ability previously well acquainted with similar or analogous subjects shou'd be able to apply these *principles* to practise and produce like effects. Justice wou'd also say, that though an inventor has given only an imperfect specification, yet if he has voluntarily instructed a number of Persons, not under ties of secrecy after the period of the patent, in the means of carrying his invention into execution, that such person did not intend to conceal the same from the publick, and as he had virtually put the publick in possession of his secret, if there was any in the case, he should not be condemn'd for an ideal crime, and a Court of Equity should in such case give him relief against the strict decision of Law. Without entering farther into this question, it is easy to see, that independant of any such condition the publick is a gainer, as it must always reap the principal part of the benefits which result from any invention that is carried on within the realm, for does not the patentee thereby become a Member of the state, if he was not so before? do not he and all his workmen pay taxes to the state and also contribute by their consumption of provisions and manufactures to support other members of it? and is not frequently the machine, or commodity he produces a foundation for the employment of thousands? The invention of Cotton spinning by Mills is a striking instance of this

As however for the general improvement of Manufactures and the diffusion of knowledge, it is to be wished that the publick shou'd at the end of the term be put in possession of the secret of the invention, if there be any in it, or of the modes of carrying it into practise *readily*, it seems proper that some regulation shou'd be made whereby these ends may be attained, and at the same time the patentee shoud be secured against having his patent overthrown for want of what may be called duly specifying –

It is therefore humbly proposed that an Act of Parliament may be made for explaining and amending the Law relating to Patents, by which it may be enacted as follows –

1st. Who are the persons entitled to patents; which appear to be the following –

The Inventors of new machines Arts and Commodities.

The improvers of old Machines Arts or Commodities, and of the method of constructing making or producing them.

Those who bring from Foreign Countries, the methods of making machines manufactures or commodities not practised or used in Britain before that time, or at that time –

Those who combine together old Instruments or machines so as to produce new effects, or to make them more extensively useful to the publick –

Those who apply old machines or instruments to new uses – Provided such new uses be essentially different from the common uses of the said machines or Instruments.

Those who by new processes produce common commodities in a better easier or more commodious manner.

No patent shou'd be valid for the sole use of any natural substance but for particular uses and applications of the same in composition or otherwise, *such uses being new.*

No patent to be valid for any improvement on a patent invention during the term of the patent unless such improvement can be practised without encroaching on the patent rights of the first Patentee.

2d. Any person thinking he has made a new discovery or invention may be empowered to give in at the office of the Attorney or Sollicitor General a Caveat against any other Persons taking surreptitiously a patent for the same, and at the same time *or at any time or times within the course of a Year* to give in sealed letters or packets explaining the then state of his ideas or experiments on the same subject, such letters to remain in the office unopened until some other application is made for a patent on the same subject. Then the Attorney or Sollicitor General to summon the parties before him, and also to summon 2 members of the Royal Society, who shall be recommended by the Council of that Society as competent Judges in the particular case, and also 2 eminent Artists in that branch of the arts to which the Caveat refers, and the Attorney or Sollicitor General presiding as judge, the said Commissioners shall open the sealed descriptions of the lodger of the Caveat, and from it and by the account of his invention given by the other party, shall determine Whether the inventions interfere, and in case the inventions are the same, or do interfere so as the one cannot be practised without encroaching upon the other, the Commissioners shall report the same to the Attorney or Sollicitor General who shall require the contending parties to produce Witnesses as to the dates of their respective inventions and shall again summon the said Commissioners to determine thereon, who shall give the cause in favour of him who is the first inventor, or of him who has brought the invention to the greatest perfection, being bona fide an inventor thereof. Or if the said parties shall have devised different means of producing the same effect then it shall be lawful for each of them to take out a Patent for their peculiar mode of proceeding – The said Commissioners and the Clerk to be bound by Oath not to divulge the secret intrusted to them for the space of one Year –

A Caveat to remain in force for one Year, when if not called in question, it may be renewed, but if not renewed, or after a decision thereon, not to come in bar of a patent, nor the sealed letter or examinations, in this Court to be adduced as evidence against a patent after the expiration of the Year.

Each of the Commissioners to receive   and the Clerk   for their trouble each time of sitting, the number of sittings not being more than 3. The expences to be defrayed equally by the contending parties – Appeal from such sentence to lye to the Courts in Westminster hall –

3d. The patent being obtained in the ordinary way the Patentee is within 4 Calendar months to deliver in a general specification of the *principles* on which his invention proceeds, and that in the most comprehensive short and clear manner which the patentee can devise; so that it may point out distinctly what is *new* and *peculiar* in his invention. Such general specifications to be preserved in the publick records of Chancery, and printed annualy, by the Kings printer, on the same size letter and price as acts of Parliament each year making a volume of all the specifications in that year but may be locked up from the public view, if the Lord Chancellor upon application made to him for that purpose shall judge it shou'd be concealed for the good of the state.

4th. Within 12 months after the date of the patent the patentee or his representatives, shall deliver in a particular description of his invention, explaining, at least, one way of putting his principles in practise, and if the said invention relate to a machine or instrument or combination or improvements of or on Machines or instruments, the inventor shall set forth the sizes shapes proportions velocities and Uses of the essential parts thereof with the materials whereof they are or may be made, illustrating the same by drawings, if need be, such drawings to be made to a scale expresst on the same, or the dimensions of the essential parts written thereon in figures, and also if any of the parts of the said invention require to be made or manufactured by any new process, he shall describe the same. But he shall not be obliged to describe particularly, the nature or method of manufacturing any such parts of the machine or invention as are in common use by others at the time, or are manufactured by Common processes, and are not intended to be secured by the Patent, in so far as regards the general use of the said parts, and in making out this description, he shall particularly note, what *parts methods* or *contrivances* he conceives to be new and of his invention, and what are only new in regard to the application thereof in this particular case – If the principles admit of being applyed in different manners and to machines of different forms so as to produce the same or similar effects, he may describe or generally mention such different modes of applying the principles as occur to him, but it shall not be deemed a defect in the specification, nor shall it empower any body to break through the patent or use any of his general principles and method of proceeding, though he shou'd not particularize such variations, provided he clearly describes one method of practising the same – And in like manner if the invention be of that sort which may be called Chemical inventions or processes, the inventor shall particularly enumerate the different ingredients which he uses, and those which may be substituted in place of them, or of any of them, the weight or measures and proportions of each of them respectively, and the method of mixing them and preparing the medicine or commodity, pointing out what are the essential articles to be attended to in every part of the process.

In inventions which are neither strictly chemical nor mechanical, or which are of a mixed nature, the same general rules to be observed in making the specifications. Within Days after the patentee shall have delivered into Chancery his said particular specification sealed up. The Master of the Rolls shall summon 5 Commissioners that is to say 3 Members of the Royal Society recommended by the Council of that society as proper persons to examine into the merits of the said specification, with 2 eminent artists skilled in the art in which the patent shall relate to, or in others analogous to it, and these 2 artists shall be chosen by the Master of the Rolls out of a list of 10 competent artists to be furnished by the patentee – And the patentee being required to attend by himself or in case of sickness, death or other legal incapacity, by his deputy or representative. The said Commissioners shall meet at such place as they shall judge proper and shall appoint a Clerk to the Commission, and the specifications both general and particular being laid before them, these Commissioners shall examine the same, and if they or the major part of them shall deem the invention to be sufficiently discriminated and explained, they shall certify the same by a writing directed to the *said Master of the Rolls*, but on the contrary, if the specifications shall not appear clear and intelligible to them, they shall interrogate the patentee, and cause their questions and his answers to be taken in writing by the Clerk to the Commission, and if they shall then be satisfied they shall report accordingly to the Master of the Rolls who shall direct the same to be enrolled in Chancery along with and annexed to the specification; But if the patentees answers to their questions shall not prove satisfactory to the said Commissioners or to the greater part of them they shall report accordingly, and the

shall direct the patentee to make out another and more perfect specification within 60 days which shall in like manner be laid before 5 Commissioners summoned as before, but if the patentee shall object at this second hearing to 2 of the 3 members of the Royal Society which have been appointed, other 2 shall be summoned in their place. If this 2nd. set of Commissioners shall not find the patentee has properly described his invention and that his answers are not satisfactory, they shall report the same and the patent shall be declared null and void – In patents for Medicines, 2 members of the Commission shall be Doctors or Licentiates in medicine and other 2 skilful working Chemists or apothecaries, and if the said Commissioners shall not judge the Medicine to be safe and wholesome it shall not be permitted to be used and the patent shall be void –

The Commissioners shall be empowered to interrogate the patentees servants, and to require the patentee to shew them the machine, instrument or contrivance which he has specified, or in place thereof a working model of the same, if the invention is of a mechanical nature, and if they shall judge such exhibition necessary, and in like manner, if the invention is an operation in the arts or in Chemistry, they may require admittance into the Laboratory or workshops where the process is carried on, or to have the operation performed before them, but it shall be in the power of the said Commissioners to dispense with such exhibition or performance of process if they are otherwise satisfied.

The Commissioners not to be empowered to judge of the merits of the invention, the novelty or Utility thereof but simply whether or not the patentee has specified the same clearly or intelligibly, but if they shall judge the same not to be a new invention, they are required to warn the patentee thereof.

The Members of the said Commission shall have a right to demand 1 Guinea each for their trouble for each time of sitting, and the Clerk

5thly. At any time or times during the term of the patent the patentee *may* give in and cause to be enrolled in Chancery specifications of such farther improvements as he may have made in the Executive part of his Invention, or on the particular forms, and proportions of the parts thereof, or the materials employed therein provided that such improvements be applicable or immediately relating to his first principles, such specifications to secure such new improvements for and during the term of his original patent, but no longer.

6thly. The *particular* specifications and post specifications to be locked up in Chancery and the same not to be shewn nor copies taken thereof without the express mandate of the Lord Chancellor which he is empowered to grant 1st. in cases of Law suits respecting the patent rights, 2dly. In Case that a subject of Great Britain residing within the realm shall make oath that he had invented a method of effecting the same thing, or nearly the same thing for which a patent had been granted to AB, and that in order to avoid interfering with the patent rights of the said Patentee and for that reason only he petitions the Lord Chancellor to grant him permission to see the specification of the said AB, such permission to be granted of course, provided that the said patentee AB has 30 days previous notice given him to attend by himself or his Attorney at the time the said specification is shewn to the said petitioner, and the said petitioner shall only be permitted to read and examine the said specification and not to take any copy or notes thereof. And after the patent is expired the particular and post specifications shall be still kept locked up, but the Lord Chancellor shall be impowered to grant copies thereof to be given to persons natives of Great Britain and residing therein petitioning for such copies and making affidavit that they intend to practise the said art or invention within the realm –

Note – In many cases the exposing the specifications and Models in a public Court may have bad consequences* by laying essential manufactures open to Foreigners, it wou'd be well therefore if the Lord Chancellor, or the Judge before whom the Cause might come to be tried were invested with a power to appoint proper persons qualified to examine into the matters of fact so far as the specification was concerned in disputes about the novelty, or prior use of an invention, which persons might be examined by the Court as evidences in matters of opinion and in the matter of fact as to the contents of the specification. But the proper arrangement of this matter, the author acknowledges to be above his abilities, it involves many difficulties. He wou'd be sorry on one hand, to see the secrets of our Artists exposed to foreigners, and on the other hand he woud be more sorry to put the decision of property into the hands of a few men who might be prejudiced or interested, he hopes however that abler heads will be able to devise some means whereby the dilemma may be avoided –

7th. As the human understanding is exceedingly fallible in respect to any nice discrimination of mechanical inventions or Chemical processes, which were performed at any distant period of time, no witness shall be examined, nor his testimony be valid concerning any facts respecting priority or novelty of invention which shall have happen'd more than 14 years before the date of his examination, or more than 5 Years before the date of the patent,

* Perhaps the most liberal way will be found the most advantageous for the publick viz. the printing all the specifications as soon as the patent is expired.

nor shall any proofs be adduced out of printed books, nor from circumstances which have taken place out of the kingdom of Great Britain, it being very plain, that if any Art has not been practised for 5 Years within the realm so that evidence of the fact can be brought the said Art was not *in esse* in so far as regards its utility to this Country and probably that it had not been brought to be useful by the first inventor thereof.

8thly. The public or common use of any invention shoud be determined 1st. By its being publickly sold in the way of trade, provided that the nature of the machine or commodity is such that the method of making it is evident from seeing and examining it – 2d. By its being erected used or practised in public places 3dly. By its being practised in open workshops 4thly. By 10 or more people, not under bonds of secrecy, having been instructed in the use practise or construction thereof – 5th. By its having been practiced in the workshops or Laboratories and by the Servants of more than one master manufacturer as a trade or business or branch thereof within one Year of the date of the patent, although such Manufacturers shou'd have kept it secret to the best of their power –

All the above to be understood of machines, instruments or processes which are or may be in *esse* or in actual practise, and not of such things as are in theory or idea. The man who reduces a theory in to practise, or recovers a forgotten art, highly merits a patent.

9th. Models of machines not publickly exhibited nor arts processes or inventions or experiments practised privately, or only shewn in confidence to the friends or servants of the Inventor, not to be esteemed prior practise, nor to operate against a patent, *unless a Caveat has been lodged*, or it can be proved that the patentee borrowed his ideas from a knowledge or hearsay of such experiments –

10th. Patents to be valid for new combinations of common machines, ingredients processes or Instruments or new uses thereof though each part shou'd have been in common use prior to the patent, provided that the effect produced by the combination or the use be new, or an improvement on the common processes. But such patent not to prevent the use of the said Common machines ingredients processes or instruments taken separately, or of such combinations or uses of them as have been commonly or previously used.

11th. No patent to be invalidated on account of the patentees specifying, (through ignorance) parts as new, which have been previously used, but such parts of the said invention as are really new to remain protected by the patent as if no such error had been committed.

12th. No patent to be set aside for want of a clear specification after the Commissioners shall have examin'd and approved of the specification –

13th. All persons now holding patents from the Crown or exclusive privileges granted by Act of Parliament for new inventions the terms of which are unexpired to be empowered if they shall judge it necessary to deliver into Chancery within 12 Calendar months of the date of the passing this Act, such further elucidations or specifications of their several inventions as they shall think proper, and . . . after the delivery the Master of the rolls shall summon Commissioners to examine the same in manner before prescribed for new patents; or upon their declaration that they are willing to abide by their original specifications, the Commissioners shall examine and report upon the same, and in drawing such new specification the patentee shall confine himself to such things as are fairly deducible from or applicable to his first principles, or are explanations thereof, or in place of making out a new specification he may refer to or elucidate the former one, and such new specification or elucidation of the old one being received and approved by the Commissioners no action at law or equity shall be brought against or pleaded in bar of his patent on account of any defect in specification. But if any patentee shall think his former specification sufficient and is willing to trust to the merits thereof in a Court of law, this Act shall not affect the validity of the same in any way, though he should neglect or refuse to submit the same to Commissioners, but all such specifications shall be and remain as if this act had never been made.

14th. In other respects all patents now in being to be judged of and determined upon according to the true intent and meaning of this act.

15th. As the general specifications being in some cases to be locked up from the public eye, persons may unwittingly offend and encroach upon the rights of patentees, therefore as soon as any patent passes the Great Seal, notice thereof shall be given in the London Gazette and also in some one or more of the evening papers which are most in repute, such notice to be twice repeated, and in cases when the general specifications are locked up in Chancery, as soon as a patentee receives authentic information of any persons encroaching upon his patent rights, he is requir'd to give notice to the parties that he is inform'd that they are so offending and requiring them to desist from the same, persons persisting after such warning to be liable to double costs and damages, but if such warning is not given, not subject to damages.

16th. No persons shall during the term of any patent right or exclusive privilege make any part or parts of the said invention or the whole thereof with an intent to send it out of the kingdom of Great Britain without the consent of the patentee, provided always, that such part, parts, ingredients or mixture are not in common use for that purpose, but are evidently applicable only to the said invention machine or process, due proof of their being in common use to rest upon the maker or exporter, and in case of any parts of inventions or Machines, or compound ingredients in chemical processes being exported out of this kingdom in order to defraud the patentee, he shall be allow'd to bring proof from Foreign Countries of the use they were there applied to, and obtain damages accordingly.

17th. No patent now in being to be set aside for a defective specification provided that the inventor shall prove that he has publickly practised the invention and that he has by himself or others instructed 5 or more persons not bound to secrecy after the term of the patent in the use of, and method of constructing making or practising the same, so that the public are effectually secured in the possession of the invention –

18th. No practise or use publick or private of any invention posterior to the date of the delivering the Affidavit and petition for a patent at the proper office to be brought in evidence against the said patent as a prior use – N.B. It is a common practise with many people to learn what petitions are presented at the offices and by exerting themselves, to procure a patent to be sealed to them before that of the original inventor or at least to create him much trouble in the course of passing his patent –

During the time of any patent for a new invention, or during the term of any exclusive privilege granted or confirm'd by act of Parliament it shall not be lawful for any person whomsoever without the consent of the patentee or of the holder of the exclusive Privilege to use or practise the said invention art or process, or to make construct, use or sell any machines instruments Utensils or Commodities made constructed or prepared according thereto, or according to any principle art method or process or any part thereof laid down in the patentees general or particular specification and thereby set forth to be appropriated and protected to the use of the patentee or holder of the exclusive privilege nor make any imitations or resemblances thereof by alterations of forms shapes positions, sizes dimensions proportions, materials or ingredients so as to pretend the same not to be according to the patentees invention.

Provided that nothing herein contain'd shall be so construed as to authorize any patentee or holder of exclusive privilege to prevent any person whomsoever from using any principle method art process ingredient instrument or utensil which has been in common use before the date of his application for his patent or from applying the same to any use or purpose whatsoever excepting always to such new uses and applications thereof as shall be of the patentees invention and set forth as such in his general or particular specification, or in the title to his patent, nor to prevent any person whomsoever from making machines instruments utensils or commodities to answer the same purposes or to produce the same effects by any principles methods arts or processes which shall not be of the patentees invention, or not set forth in his specifications and not deducible from or parts of these contain'd in the said specification.

Persons imitating making using or selling patent inventions before the date of delivery of general specifications of patents not liable to damages, provided they desist at the said date.

Priority of the dates of patents to be reckon'd from the dates of the application for them respectively and not by the dates of the affixing the Great Seal, provided that no unnecessary delay in solliciting the patent arise on the part of the Petitioner.

Note – In framing that part of the Law which regards inventions practised in foreign Countries regard shou'd be had to the following circumstances. On the one hand, it woud be hard to deprive an ingenious foreigner of the benefits which he might reap from his invention provided he chose to practise it here; or to confine the artists of this Country by giving any person other than the inventor a patent for what he might have no merit in, farther than being the 1st. practiser in this Country, and which he might not be able to make so useful as the inventor could: On the other hand, if the bringing an Art from a foreign Country is attended with expence risk and trouble, the person who first introduces it in Britain deserves a patent – In order therefore to give the 1st. and real inventor a fair chance, and also to give this Country a chance of obtaining the invention without the burthen of a patent, or if that does not happen in a reasonable time, to encourage Artists to bring inventions or arts from foreign Countries, by securing to them such profits as may arise from a Patent, suppose it were enacted, that 'for the space of 1 Year from the date of the '1st. public use of any new invention in foreign Countries, no person except *the true inventor* '*thereof* shall have a right to take a patent for the same in Britain. But that provided the 'invention or art has not been previously practised, or the commodity or article made or 'manufactured in Britain, the inventor shall have a right to ask a patent for the said art 'manufacture machine or invention, either within the aforesaid term of 1 Year from the 'date of the 1st. public use thereof in Foreign Countries or at any time afterwards previous 'to the publick practise by others of the said art, invention etc. in Great Britain But if the 'inventor shall not apply for a patent within the aforesaid term of 1 Year from the date of the '1st. public use of the said invention, and the same shall not be or have been publickly 'practised by any person or persons within Great Britain so as to be useful or successful, 'then and in every such case it shall be lawful to and for any person whomsoever who shall 'have learn't acquir'd or brought from foreign Countries the Art of constructing, making or 'manufacturing any Machine commodity Utensil or other invention whatsoever to apply 'for a patent for the same, which shall be good and valid, and to be judg'd of in every other 'respect as if the said first practiser of the invention in Britain had been the first and true 'inventor thereof, and in regard that machines or commodities may be used in Britain the 'art of making or manufacturing which is not understood, or may not be practised in 'Britain, be it enacted, that the publick practise of any invention which shall operate against 'the validity of a patent for the same, shall be understood to be confin'd to the public practise '*of the making and manufacturing the same*, and not to the use thereof, though such use shall 'be common at the time of granting the patent' –

The want of determinate laws, ascertaining the duties and rights of Patentees is much felt, were they amended, or at least made more explicit, the number of law suits on these subjects would be fewer, and Men of ingenuity wou'd be better employ'd in finding out new arts than in endeavouring to deprive their Neighbor of the benefit of his inventions – It may perhaps be objected that the times proposed for giving in the general and particular descriptions are too long. Many inventions cannot be fairly tried but in a public manner, therefore the patentee shou'd have time to make some experiments before he shoud be obligd to give any account of his invention particularly, as the general specification shoud be so well drawn that no man can find any flaw in it, whereby he may evade the patent, and accidents of bad health, or necessary attendance on other business, may if confined to a term render it impossible for the patentee to do himself Justice. And as to the *particular* specification, as the inventors pretensions are regulated and defined by the general one, it seems that it shoud be proper to give him time to make improvements on his first ideas, particularly in what regards the Minutiae of the parts of the Machine or process, that the public may be put in possession of his invention in a more full and complete manner than they cou'd otherwise be. For altho' it may be said that an inventor shou'd have fully ascertain'd his machine or process before he demands a patent, yet every body who has been conversant in the arts must know that most machines or contrivances however simple they may appear cannot be brought to perfection without some experience, and that experience cannot be attaind without trusting many people with the secret –

The frequency of Patents and the frivolity of many of the inventions are complaind of, but such patents do little if any harm to the Community, and it will be difficult to contrive any means whereby valuable inventions may be distinguished *a priori* from those that are not so, any Court of enquiry into their merits which cou'd be instituted might form false judgments, genius might be repressed or deterr'd by difficulty and expence from pursuing its schemes, and an inventors secret might be wrested from him and a patent refus'd because Jurymen were wrongheaded, did not understand the subject, or were interested in the consequences. It seems therefore essential that the inventor be allow'd to keep his secret until he is secur'd by the patent. This may be well illustrated by the example in Holland, where tho' the fees on a patent are very trifling, it costs more time and trouble, more making of interest with those in power, more proof of Utility, and more negotiation to obtain a patent, then woud suffice to make a treaty of peace between 2 contending Nations. The consequence is, that Country produces few new inventions or improvements in the Arts, now a days, tho' formerly it surpassed all Europe as a manufacturing Country –

Though the object of a patent seem trifling, it sho'd not be despised, for every trifle which is amusing, or useful produces a new branch of trade; a new method of making a button may make the fortune of the inventor and add riches to the state. And even granting that the greater number of Patents shou'd prove useless both to the state and to the inventors, *if one Capital and meritorious invention such as Sir Richard Arkwrights Cotton machine should be brought forth in a Century in consequence of them* it shou'd justify the measure of granting them. But Sir Richards is far from being the only instance, several very important, ingenious and useful inventions have been brought to perfection in consequence of patents now existing, and more may be expected, when patentees have a better security given for their property.

But if patentees are to be regarded by the public, or held up to it in the odious light of monopolists, and their patents considered as nuisances and encroachments on the natural liberties of his Majesty's other subjects, if their acquiring fortunes by their ingenuity is to be considered as a crime, if they are to be deprived of their rights as soon as it appears that they have made an advantageous or lucrative discovery, because they have not specified in a manner which the patent did not require of them, and which if it had been required it might not have been in their power to fulfill, without there had been some officer appointed, who shou'd have told them, 'we cannot receive your specification without' you will make it more clear in such and such points; If combinations are permitted to be formd to harrass them with law suits, and if in case of a trial with an invader of their patent rights, a person who perhaps sets up no pretensions to any other merit than his being bold enough to vindicate in a Court of law, the wrongs he had done the patentee, if in such case, every evasion of the law is to be employ'd to shelter their opponent and every piccadillo magnified against the patentee, wou'd it not be just to make a law at once, taking away the power of granting patents for new inventions and by cutting off the hopes of ingenious men oblige them either to go on in the way of their fathers, and not spend their time which sho'd be devoted to the encrease of their own fortunes in making improvements for an ungrateful public, or else to emigrate to some other Country that will afford to their inventions, the protection they may merit. But the author cannot doubt, that those who manage the public affairs are too sensible of the advantages which the nation reaps from its manufactures, of the great consequence that improvements in the arts are to their prosperity and of the justice of protecting men of ingenuity, to hesitate on the line of conduct they will pursue, when once they are satisfied as to the form of a law that will answer These Ends.

Engineers' wash drawings relating to the Patents of 1781 (see page 23), 1782 (see page 26) & 1784 (see page 29)

Drawing N°1, relative to the first method.

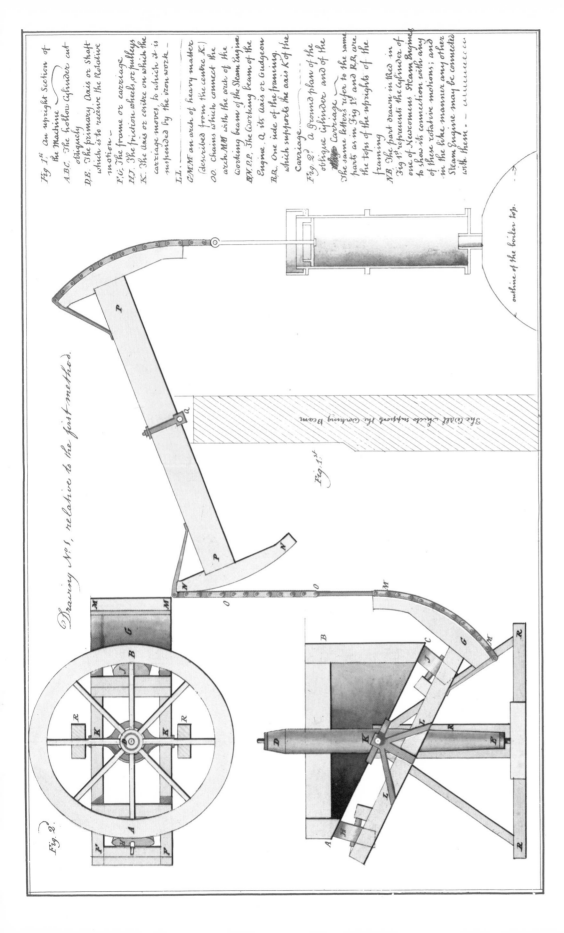

Fig 1st. An upright Section of the Machine

A.B.C. The hollow Cylinder cut obliquely

D.E. The primary Axis or shaft which is to receive the Rotative motion —

F.G. The frame or carriage

H.J. The friction wheels, or pulleys

K. The Axis or centre on which the carriage moves, to which it is suspended by the iron work. —

I.I.

G.M.M. an arch of heavy matter (described from the centre K)

O.O. Chains which connect the arch M.M with the axis of the working beam of the Steam Engine

N.N.D.P. The working beam of the Engine. Q. its Axis or Gudgeon

R.R. One side of the framing which supports the axis K of the Carriage.

Fig 2d. A ground plan of the oblique cylinder and of the Carriage.

The same Letters refer to the same parts as in Fig 1st and R.R are the tops of the uprights of the framing

N.B. The part drawn in Red in Fig 1st represents the cylinder of one of Newcomen's Steam Engines, to shew its connection with any of these rotative motions; and in the like manner any other Steam Engine may be connected with them. — แแแแแแแ

Fig. 1st

The Wall which supports the working Beam

outline of the boiler top.

Fig 2d.

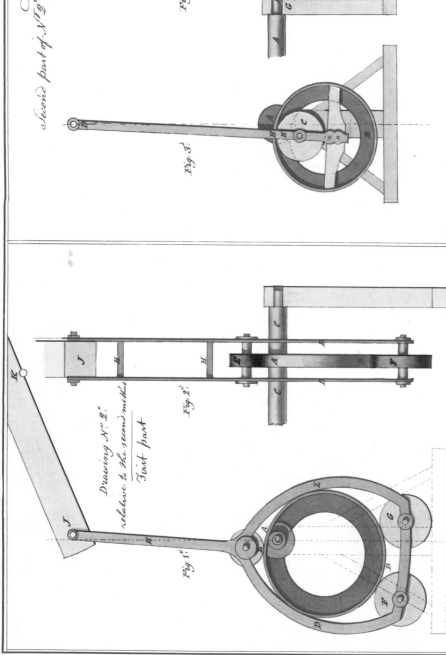

1781

Second part of N.º 2

Fig. 3.ᵈ

Fig. 4.

N13 Fig 4.ᵗʰ is only an edge view of the friction wheel C, and the connecting Rod ED, But is a Section of the Excentric wheel AB.

Fig 3.ᵈ & 4.ᵗʰ Represent the application of the second method to the internal cir-cumference of an Eccentric wheel ~
Fig 3.ᵈ View of the excentric wheel in the direction of its shaft. A the end of the axis or shaft, which is behind the wheel. AB the Excentric wheel. AB the Excentric wheel ~
C, The friction wheel, which is connected with the working beam of the Steam Engine by means of the rod ED, which is suspended to the working beam by the end D
Fig 4.ᵗʰ. Edge view of the excentric wheel and its shaft &c~

Drawing N.º 2:
relative to the second method
First part

Fig 1.ˢᵗ

Fig 2.ᵈ

Fig 1.ˢᵗ A view of a circular Eccentric wheel seen in the direction of its shaft or axis AB. The excentric wheel. C. Its Shaft or axis. EFG. Friction wheels.
HDF, a frame of iron by which the friction wheels are suspended to and con-nected with JK, one end of the working beam of the Steam Engine ~
Fig 2.ᵈ An edge view of the excentric wheel ~ The same letters relate to the same parts as in Fig 1.ˢᵗ

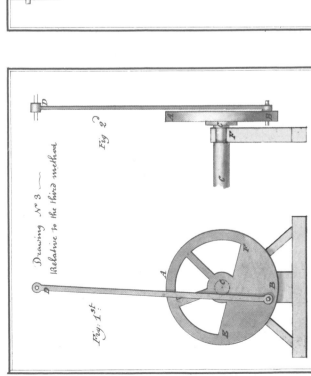
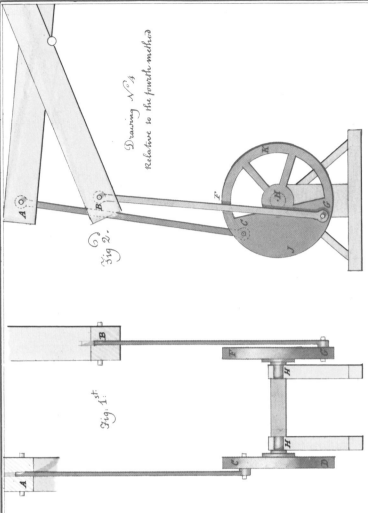

Drawing Nº 5
Relative to the fourth method

Fig. 2.

Fig. 1st.

Fig 1st. An Edge view of the two Wheels and a longitudinal one of their Shaft or axis
A and B parts of the working Beams of the two Steam Engines. C.D.F.G Edge views of the wheels
H.H. Their common axis supported at H H. — A.C,B.G Rods which connect the ~~following~~ wheels
to the working beams of their respective Engines.
Fig 2ª. Front view of the wheel F G (the other wheel C.D. being hid by F.G, in this view) —
A.B. Xart of the Working beams of the two Steam Engines. A C The connecting rod of the further
Wheel C.D. — B,G, the connecting rod of the nearest wheel (J F.G.)
H. The Centre of the wheels and of their Shaft or axis. — C J G The heavy side of both wheels.

Drawing Nº 3 —
Relative to the third method.

Fig 2

Fig. 1st.

Fig 1st. Front view of the primary Rotative wheel
A.B. The Wheel,   C. its centre, or axis
E.B.F. The heavy side of the wheel. B The point of
attachment of the rod B.D which connects the wheel
with the Working Beam of the Steam Engine.
D The end of the rod which is attached to the working beam
Fig 2ª. An edge view of the wheel and its shaft &c

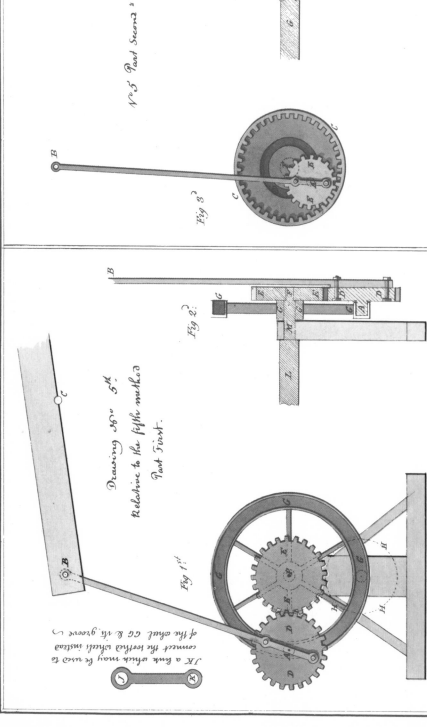

N.º 5 "Part Second"

Fig 4.

Fig 3.

Fig 3.ᵈ Front view of the application of the Fifth method to a Rotative wheel having teeth on its internal circumference – CC The Rotative wheel which is made to turn round by the action of the wheel EE within it. EE a wheel fixed to the connecting rod AB. The pin A keeps the two wheels always in contact by moving in the groove DD.

Fig 4.ᵗʰ Section of the machine through its axis

Drawing N.º 5.ᵗʰ
Relative to the fifth method
Part First.

Fig 1.ˢᵗ

Fig 2.ᵈ

JK a link which may be used to connect the teethed wheels instead of the wheel CC & th groove

Fig: 1.ˢᵗ Front view of the Machine. – B Part of the Working beam of the Steam Engine, C its Gudgeon or axis, DD, a toothed wheel fixed to the connecting rod AB which without turning upon its own centre revolves round the other toothed wheel EE which is fixed upon the primary Rotative axis F, and which is made to turn round by the teeth of DD. – A a pin projecting from the back side of DD, which being guided by the groove GG keeps the toothed wheels always in contact – Fig 2.ᵈ Edge view of the Machine

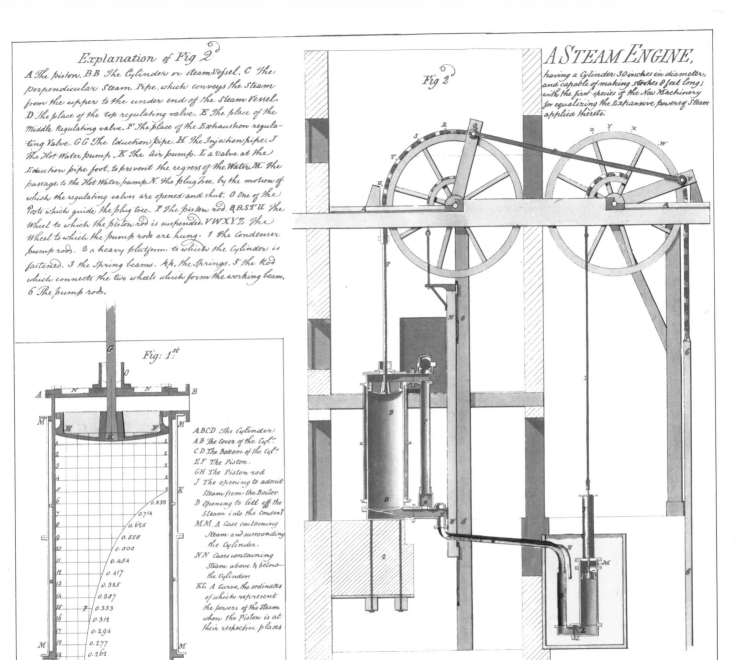

Explanation of Fig 2.

A The Piston. B B The Cylinder or steam Vessel. C The perpendicular Steam Pipe, which conveys the Steam from the upper to the under end of the Steam Vessel. D The place of the top regulating valve. E The place of the Middle Regulating valve. F The place of the Exhaustion regulating Valve. G G The Eduction pipe. H The Injection pipe. J The Hot Water pump. K The Air pump. I a Valve at the Eduction pipe foot, to prevent the regress of the Water. M The passage to the Hot Water pump. N The plug tree, by the motion of which the regulating valves are opened and shut. O One of the Parts which guide the plug tree. P The piston rod. QRSTU The Wheel to which the piston rod is suspended. VWXYZ The Wheel to which the pump rods are hung. 1 The Condenser pump rods. 2 a heavy platform to which the Cylinder is fastened. 3 the Spring beams. 4,4, the Springs. 5 the Rod which connects the two wheels which form the working beam. 6 The pump rods.

Fig. 2.

A STEAM ENGINE,

having a Cylinder 30 inches in diameter, and capable of making strokes 8 feet long; with the first species of the New Machinery for equalizing the Expansive powers of Steam applied thereto.

Fig: 1st

ABCD The Cylinder
A B The Cover of the Cyl.r
C D The Bottom of the Cyl.r
E F The Piston.
G H The Piston rod
J The opening to admit Steam from the Boiler.
D Opening to let off the Steam into the Condens.r
M M A Case containing Steam and surrounding the Cylinder.
N N Cases containing Steam above & below the Cylinder.
K L A Curve, the ordinates of which represent the powers of the Steam when the Piston is at their respective places

0.833
0.714
0.625
0.555
0.500
0.454
0.417
0.385
0.357
0.333
0.312
0.294
0.277
0.262
0.250

1782

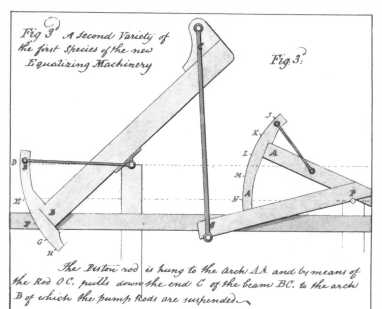

Fig. 3. A Second Variety of the first Species of the new Equalizing Machinery

Fig. 3.

The Piston rod is hung to the Arch AB and by means of the Rod OC pulls down the end C of the beam BC to the arch B of which the pump Rods are suspended

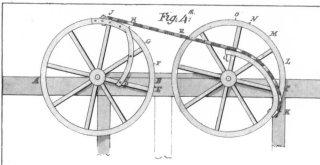

Fig. 4th The Second Species of the New Machinery for Equalizing the Expansive Powers of Steam

The Piston of the Steam Vessel is hung by chains, or otherwise to the side A of the wheel AB and by means of the chain JRK and the spiral wheels JP & QK pulls round the wheel DC to which the pump Rods are suspended, from the Arch OC.

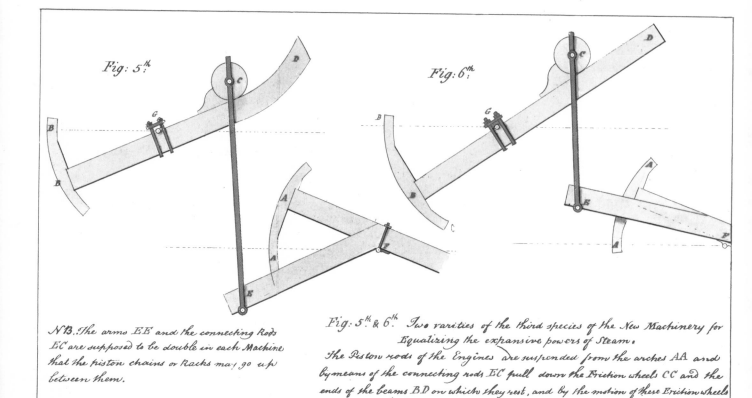

Fig. 5th

Fig. 6th

N.B. The arms EE and the connecting Rods EC are supposed to be double in each Machine that the piston chains or Racks may go up between them.

Fig. 5th & 6th Two varieties of the third species of the New Machinery for Equalizing the expansive powers of Steam.

The Piston rods of the Engines are suspended from the arches AA and by means of the connecting rods EC pull down the Friction wheels CC and the ends of the beams BD on which they rest, and by the motion of these Friction wheels on ye Beams the levers are lengthned nearly as the powers of the steam diminish.

1782

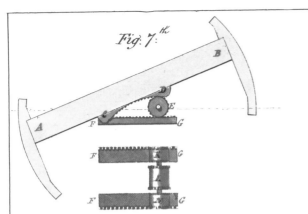

Fig: 7.th

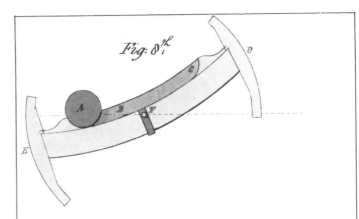

Fig: 8.th

Fig 7th. The Fourth Species of the new Equalizing Machinery
AB The Working Beam, CD A hollow curve which rests
upon the roller E and is furnished with teeth to keep
it from sliding. FG. A Flat Platform on which the end
Rollers rest. K.L.M. Is a horizontal view of the Rollers and
Platform. K&M are fast on the axis & L can turn round on it.

Fig. 8th The first Variety of the Fifth Species of the New
Equalizing Machinery.
ED The Working beam. A a weight which rolls in
a hollow curve BC. — F the axis or Gudgeon

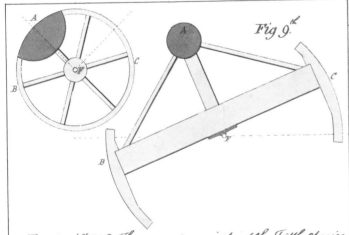

Fig: 9.th

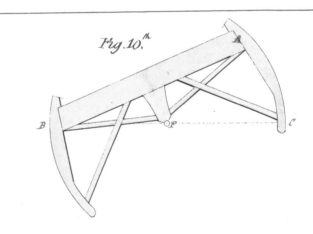

Fig. 10.th

Fig: 9. Nos 1&2 The second variety of the Fifth Species
of the new equalizing machinery. in Two Examples.
AA The heavy Bodies placed or fixed above the Centres
of motion of the Wheel BC, No 1 and of the Working beam
BC, No 2 — FF The Centres of motion.

Fig 10th. The Third Variety of the Fifth Species of the
New Equalizing Machinery
AB The Working beam, which is placed so high
above its centre of motion F that its own gravity acts the part of the weight
A in Fig. 9.

1782

Fig: 11.th The Fourth Variety of the Fifth Species of the New Equalizing Machinery.
AA.BB Two Cylinders containing water and connected by the Trough (FF). Their Pistons
C & D are supposed to be suspended from the opposite ends of a Working Beam whose
centre of motion, or Gudgeon. is supported upon the Wall EE and which beam is
so connected with the Piston of the Engine, that when the said piston descends it
raises the piston of the Water Cylinder, which is then lowest, & thereby causes the water
which it contains to pass by the trough into the other Water Cylinder, and by its
gravity to assist the descent of the Piston of the Engine.

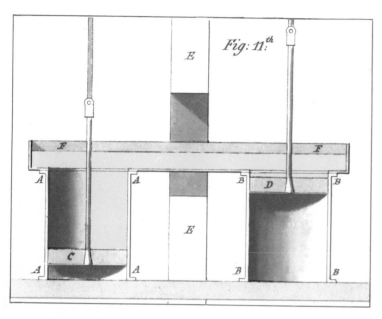

Fig: 11.th

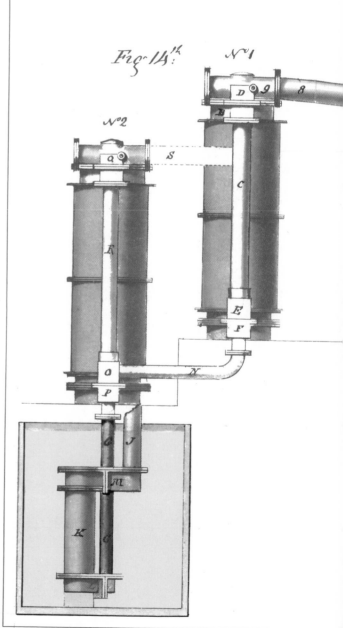

Fig: 14.th

Nº1

Nº2

Fig: 14.th Represents a Front view of the New Compound
or Double Engine as seen from the Lever Wall

Nº1. The Steam Vessel and some other parts of the Primary
Engine. Nº2. The Steam Vessel and other parts of the
Secondary Engine. D.Q. Places of the top Regulating
Valves. C.R. The perpendicular Steam pipes. E.O. Places
of the Middle Regulators. F.F places of the Exhaustion
Regulators. N.G Eduction pipes. K.J Air & Hot water
pumps. M. Passage from the Air pump to the Hot water pump.
S a pipe for communicating the Steam from the Primary
Engine to the secondary Engine, instead of the Eduction
pipe N. 8,9 Pipes which convey the Steam from the Boiler
N.B. These Engines may each have a Condenser or the same
Condenser may serve them both, as is here delineated.
    The side view of these Engines would appear the
same as Fig: 2ª or Fig. 12.th according to the construction
of their Working beams. The Hot Water pump is broken off
in this drawing to avoid its interfering with the Eduction
pipe N

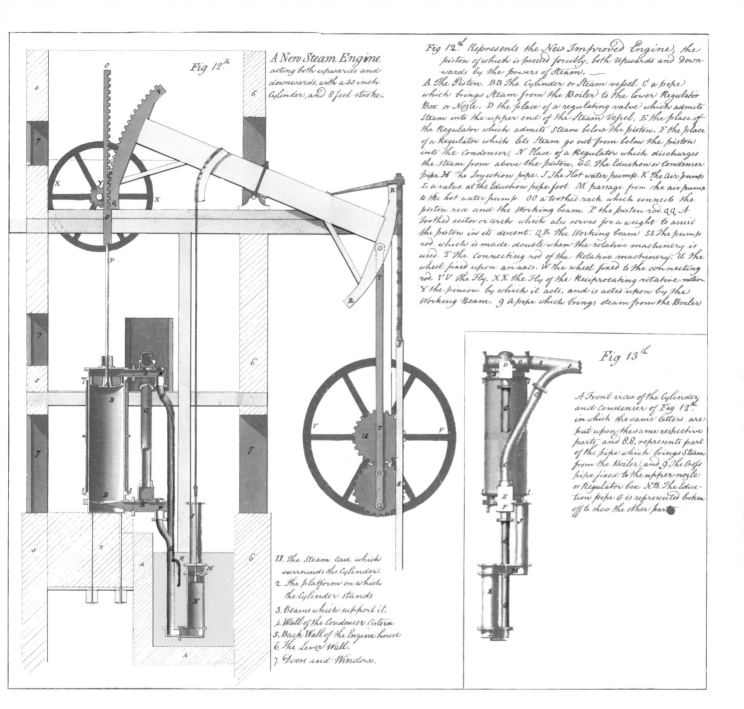

Fig 12th

A New Steam Engine
acting both upwards and
downwards, with a 30 inch
Cylinder, and 8 feet stroke.

Fig 12th. Represents the New Improved Engine, the
piston of which is pressed forcibly, both Upwards and Downwards by the powers of Steam. —
A The Piston. BB The Cylinder or Steam vessel. C a pipe
which brings Steam from the Boiler to the lower Regulator
Box or Nozle. D the place of a regulating valve which admits
Steam into the upper end of the Steam Vessel, E the place of
the Regulator which admits Steam below the piston. F the place
of a Regulator which lets Steam go out from below the piston
into the Condenser, N Place of a Regulator which discharges
the Steam from above the piston. GG The Eduction or Condenser
pipe. H The Injection pipe. I The Hot water pump. K The Air pump
L a valve at the Eduction pipe foot. M passage from the air pump
to the hot water pump. OO a toothed rack which connects the
piston rod and the Working beam. P the piston rod. QQ A
Toothed sector or arch which also serves for a weight to assist
the piston in its descent. QR The Working beam SS The pump
rod, which is made double when the rotative machinery is
used. T The connecting rod of the Rotative machinery. U The
wheel fixed upon an axis. W the wheel fixed to the connecting
rod. VV The Fly, XX The Fly of the Reciprocating rotative motion
Y the pinion by which it acts, and is acted upon by the
Working Beam. 9 A pipe which brings steam from the Boiler

Fig 13th

A Front view of the Cylinder
and Condenser of Fig 12th
in which the same letters are
put upon the same respective
parts, and 8.8. represents part
of the pipe which brings Steam
from the Boiler, and 9 The Cross
pipe fixed to the upper nozle
or Regulator box N.B. The Eduction pipe G is represented broken
off to shew the other parts

11. The Steam case which
  surrounds the Cylinder.
2. The platform on which
  the Cylinder stands
3. Beams which support it.
4. Wall of the Condenser Cistern
5. Back Wall of the Engine house
6. The Lever Wall.
7. Doors and Windows.

1782

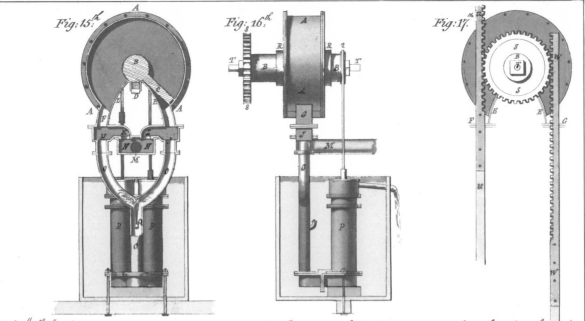

Fig: 15.ᵗʰ   Fig:ˢ 16.ᵗʰ   Fig: 17.ᵗʰ

Fig.ˢ 15.ᵗʰ 16.ᵗʰ 17.ᵗʰ. The New Reciprocating, Semi Rotative Engine. Fig 15.ᵗʰ Section of the Engine at right angles to the axle of the Engine Cylinder. AA The hollow Cylinder cut open. B. The axle or axis. C The Piston. D A Box filled with some soft Substance to make the joining of the division plates EE with the axle, steam and air tight. FG pipes which admit and discharge the Steam. HKLI places of valves or Regulators. M The Steam pipe from the Boiler. NN The Steam regulator box. OO The Eduction or Condenser pipe. Q the injection pipe. PP Condenser pumps. Fig 16.ᵗʰ a side view of the Engine wherein the same letters are placed on the same parts as, in Fig 15.ᵗʰ and RR are sockets to make the axle air tight & SS The wheel which acts upon the pump rods. 2 the wheel which works the Condensers Fig 17. Outside Front View of the Engine & pump rods, The Condenser & Regulator Boxes are not drawn in this view &, the upper part of the pump rod UU is supposed broken off. T The pivot of the axle. UU. WW. The pump rods and their racks.

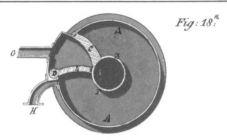

Fig: 18.ᵗʰ

Fig 18.ᵗʰ A New Rotative Engine

The Steam enters by the pipe G & acts against the Valve E and the moveable radius or piston C and the Space AABB being exhausted, the Piston revolves through it by the action of the steam, and turns the axle BB, when the piston comes to the valve E the pipe G is shut and the valve E opens by turning upon the joint D & so permits the piston C to pass by it. The Steam then rushes into the condenser by the pipe H, which exhausts the Steam Vessel. When the piston has got to its original place, the valve E is again shut and the steam admitted through the pipe G into the space between ẏ valve E & the piston C, and to continue the motion during the time that the piston is passing ẏ valve a heavy fly is fixed on or connected with, some part of the axle BB on the outside of the Cylinder

The Outside Figure of this Engine is nearly the same as Fig 16.ᵗʰ But the steam vessel may be placed either vertically, as drawn, or horizontally, as its use may require.

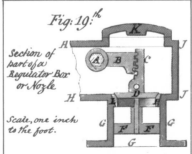

Fig: 19.ᵗʰ

Section of part of a Regulator Box or Nozle

Scale, one inch to the foot.

A Cross Section of a Spindle which comes through the side of the Box and moves the arm or Sector B which acts upon the Rack C which raises the Regulating valve D which is ground to fit the seat EE and is guided by the socket FF — GG the pipe which leads to the condenser — K A Cover which is opened occasionally to rectify the Valve HH— II part of the Nozle. —

1782

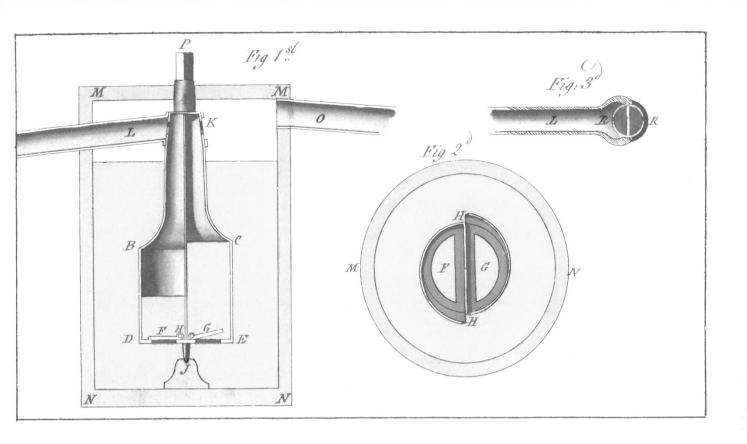

Fig 1.ˢᵗ

Fig: 3.ᵈ

Fig 2.ᵈ

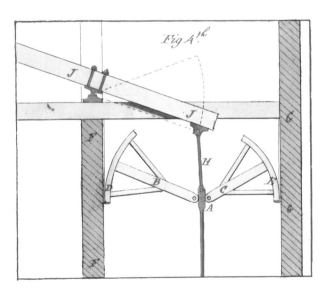

Fig 4.ᵗʰ

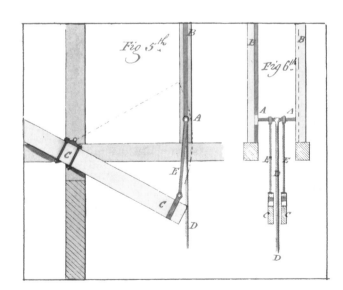

Fig 5.ᵗʰ

Fig 6.ᵗʰ

1784

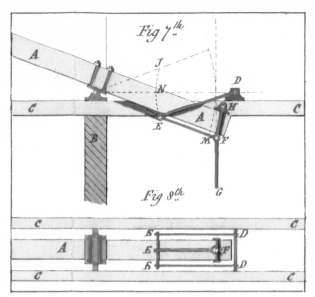

Fig 7th.

Fig 8th.

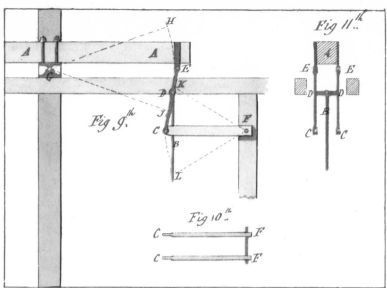

Fig 9th.

Fig 10th.

Fig 11th.

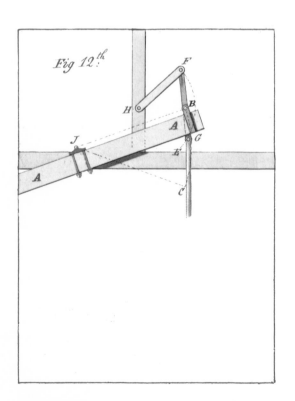

Fig 12th.

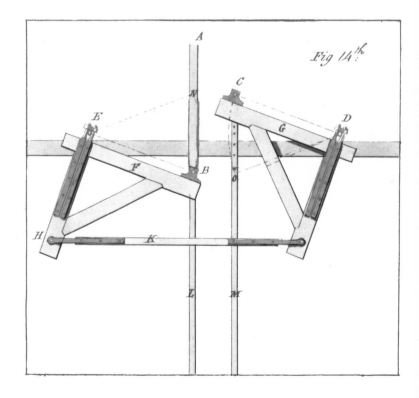

Fig 14th.

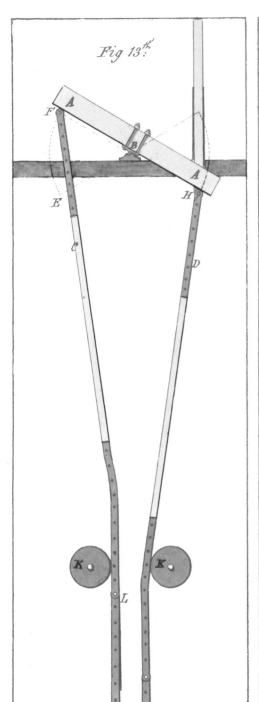

Fig 13.th

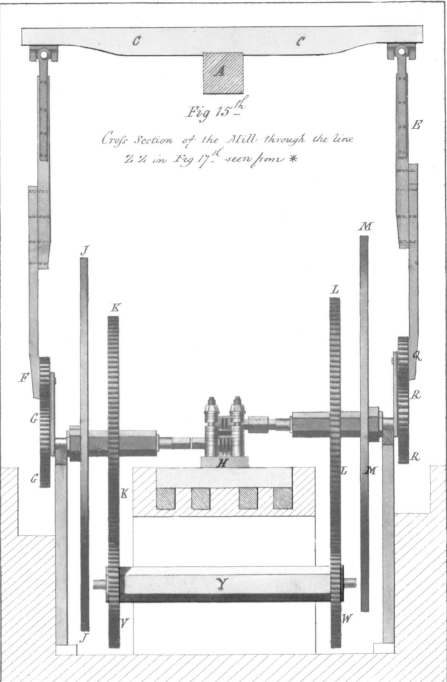

Fig 15.th

Cross Section of the Mill through the line
Z Z in Fig 17.th seen from *

1784

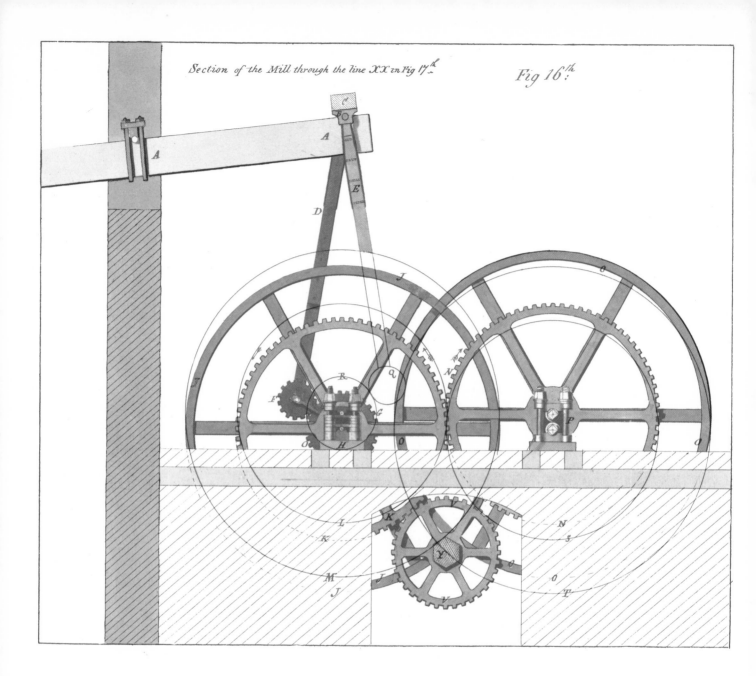

Section of the Mill through the line XX in Fig 17th.

Fig 16th:

1784

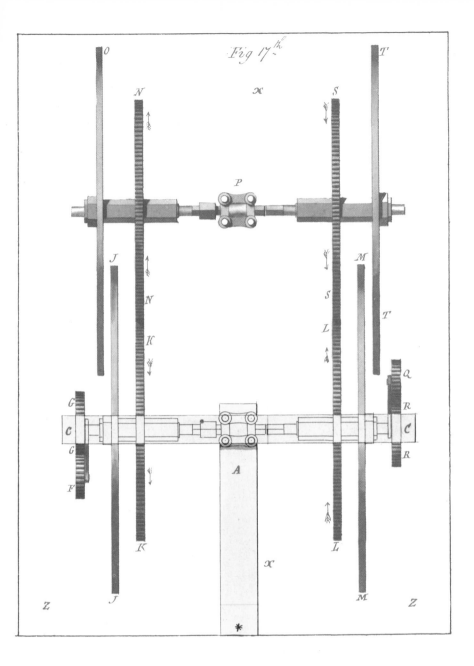

Fig 17.th

1784

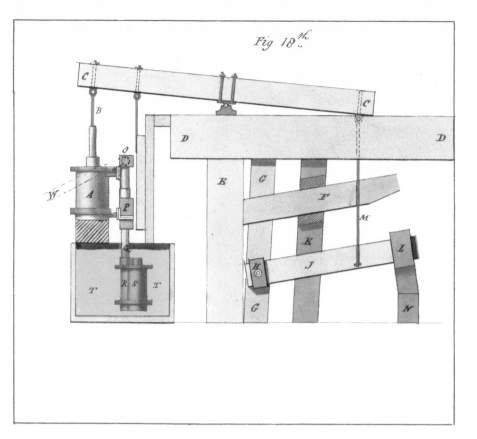

Fig 18.th

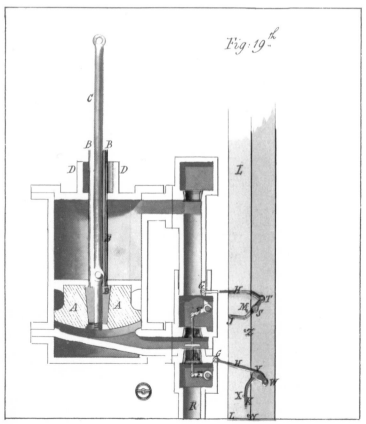

Fig: 19.th

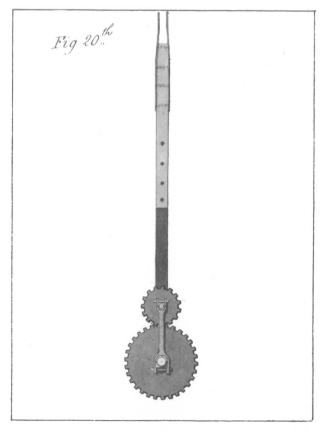

Fig 20.th